THE
TENACIOUS DREAMS
OF
GEORGE WASHINGTON CARVER

Geo. W. Carver

Carolyn T. Reeves, Ed. D.

HENRY LYON BOOKS

Published by Henry Lyon Books
an imprint of Master Design Marketing, LLC
30 N Gould St, Ste R, Sheridan, WY 82801
HenryLyonBooks.com

Cover design by Rose Carman
Interior design by Faithe Thomas

Photos/Images:
Unless otherwise noted, images are from the Public Domain.
Cover icons: Peanut by Tatyana, pocket knife by teleymon, painting by Kieu Thi Kim Cuong, graduation by Cuby Design, plant testing by Vectors Point, Thought Bubble by Abdo, Thought Bubble by Adriano Emerick from the Noun Project. Used with permission.
Chapter 1 Map: This image was originally posted to Flickr by NOAA Photo Library at https://flickr.com/photos/51647007@N08/9716190703 (archive). It was reviewed on 23 July 2018 by FlickreviewR 2 and was confirmed to be licensed under the terms of the cc-by-2.0.
Chapter 2 Second image was originally posted to Flickr by Osbornb at https://flickr.com/photos/12547641@N00/7100120493 (archive). It was reviewed on 10 December 2019 by FlickreviewR 2 and was confirmed to be licensed under the terms of the cc-by-2.0.
Chapter 3 Image used with permission. https://digitalcollections.nypl.org/items/510d47df-a272-a3d9-e040-e00a18064a99
Chapter 7 Image from https://www.flickr.com/photos/luigi_and_linda/3835814423
Chapter 12 The image has been restored, with dust spots removed, and with levels adjusted to better match his actual skin tone. It was quite common for photographers to overexpose images of darker-skinned people to bring out details, but this was often fixed for publication, and, if uncorrected, has problematic consequences for how history is imagined.
Chapter 12 Image by Charli Kendricks

Print ISBN: 978-1-947482-37-1
Ebook ISBN: 978-1-947482-38-8

Printed in the USA

TABLE OF CONTENTS

NOTE TO TEACHERS

The story of George Washington Carver takes place between the Civil War and the beginning of World War II. A brief postscript section continues into the Civil Rights era and on to current events. The book allows students to examine the disastrous effects of the philosophy of white supremacy on black citizens, as well as to trace its foundational beginnings. Students will be able to trace the restoration of civil rights to all citizens as guaranteed by the U.S. Constitution.

This book can be added to many related science or history curriculum without disrupting the existing content. It will enrich students' knowledge about this important period of history and help them make connections that may not be obvious otherwise.

It can be a stand-alone homeschool/Christian school unit or a parent-selected book to help students learn about a difficult period in American history. It can be a partner to another book in this series— *The Tenacious Dreams of Lise Meitner*, the story of a Jew who lived in Austria and Germany from 1878 to 1968. Both George and Lise lived through a period of cruel racial laws. Both were victims of discrimination because of widespread beliefs about inferior and superior races. The similarities in how the lives of both were affected by racial lies are worth examining. It is encouraging to see how both Lise and George were able to overcome the obstacles in their lives and achieve their childhood dreams.

Teacher discretion should determine whether to have students orally discuss the review questions or write their answers. Many of the topics are directly related to issues and problems in today's culture. Students are asked to consider the value or negative influences of the newer ideas being taught in Critical Race Theory and in other racial philosophies.

PROLOGUE

George Washington Carver is one of the heroes of American history. He helped many farmers, white and black, to escape deep poverty and survive, as they faced the dark days of the Reconstruction, a boll weevil invasion, and the Depression of 1929, and then go on to make farming a prosperous career.

His life began as an orphan, whose slave mother was kidnapped by a ruthless gang of bushwhackers. His surrogate parents were a white couple, who nursed him through a serious illness as a baby and taught him the value of a strong work ethic.

George's desire to learn to read and get a real education carried him through a series of great difficulties and adventures. His early morning conversations with "Mr. Creator" gave him the tenacity to finish his dream of obtaining an education and then to tackle another dream to help poor southern farmers learn better methods of farming.

He discovered more than 300 uses for peanuts and other legumes that replaced cotton as the only cash crop some farmers knew how to plant. But his most valuable legacy was the young people he taught for decades at Tuskegee Institute.

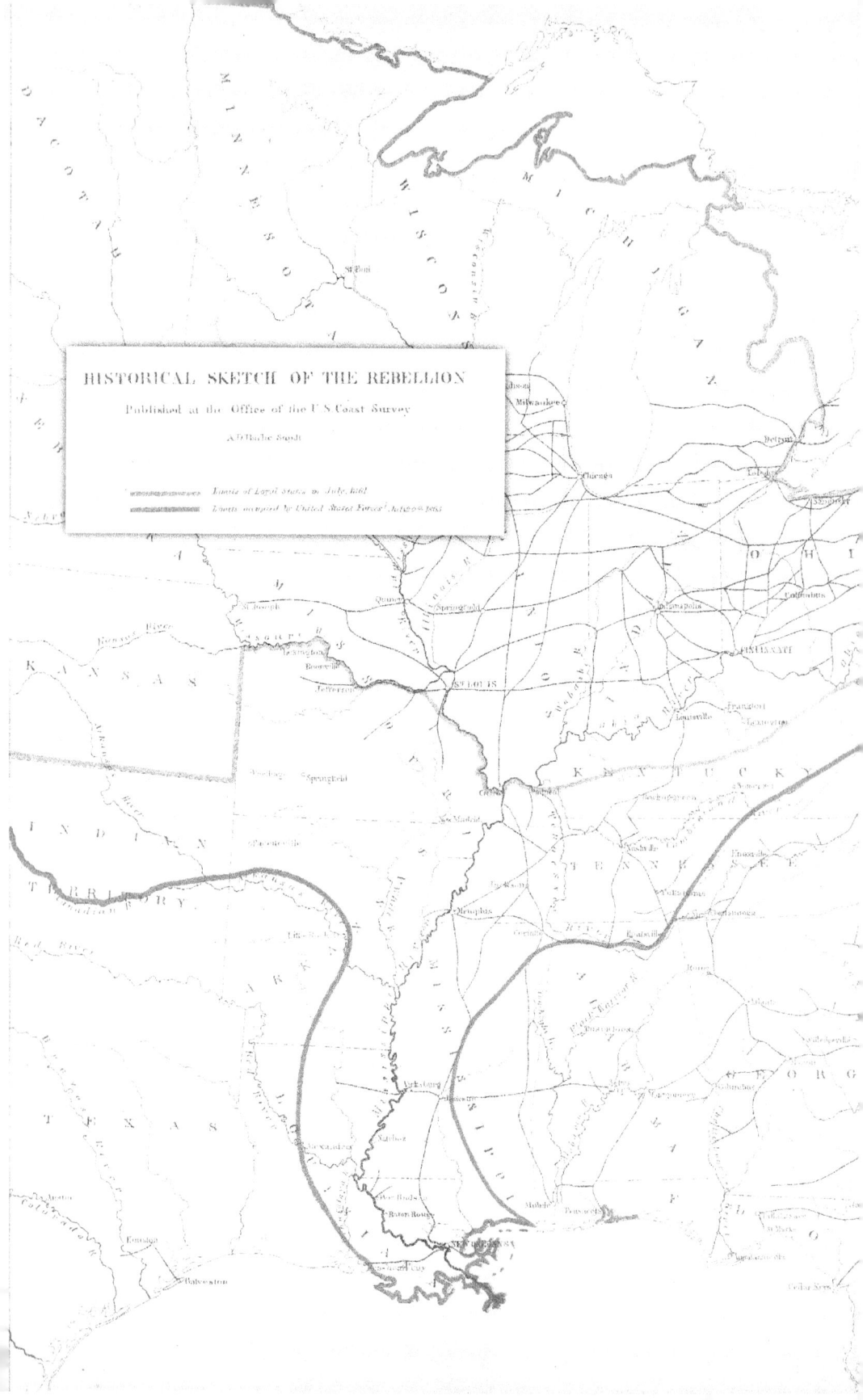

HISTORICAL SKETCH OF THE REBELLION

Published at the Office of the U S Coast Survey

A.D. Bache Supdt

Limits of Loyal States in July, 1861

Limits occupied by United States Forces between 1861

CHAPTER 1

FATHERLESS AND MOTHERLESS

Hard working and self-sufficient, Moses and Susan Carver initially thought their 240-acre farm in southwestern Missouri was out of reach of the problems dividing the nation. But before they knew what was happening, an outright Civil War loomed in their future. Located between Kansas who sided with the Union sentiments and Arkansas who sided with Confederates, Missouri tried to maintain a neutral position. As fighting between the two sides began in 1861, remaining neutral was not an easy option. The Carvers tried to continue farming as they always had, but with an unstable government, law and order was sporadic at best. This left the Carver farm right in the middle of one of the most lawless places in all of North America.

Gangs of raiders, known as bushwhackers, took advantage of these conditions and roamed the area with few restraints as they stole whatever they found of value. Bushwhackers had already raided the Carver farm three times. The once tried unsuccessfully to force Moses to tell them where he had hidden some money by branding the bottom of his bare feet with red hot coals. So far, the Carvers had been able to hold on to their possessions.

When tensions leading up to the American Civil War began to heighten, the Carvers realized they were not going to be able to keep up with all the work their farm required. Both Moses and Susan sided with the abolitionists and opposed the principles of slavery, but they desperately needed help on their farm. With able-bodied men leaving the neighborhood, they purchased a young slave named Mary for $700 to help on the farm.

Mary was always treated kindly, and the relationship between Mary and the Carvers became one of genuine friendship. During her time with the Carvers, Mary gave birth to George, as well as an older brother, James, and twin girls, who died as infants. The children's father was a slave who belonged to the owner of the adjoining farm. He was killed shortly after George was born while hauling logs to town in an ox wagon.

The Civil War had been raging for three years in 1864 when George was born. His health had not been good from the beginning, and Mary feared he would be another casualty like his twin sisters.

So far, no major battles had been fought in Missouri, but the lawlessness surrounding them was a constant danger. Everyone expected Mary would be free to leave as soon as the war ended. For now, she and her sons felt secure on the Carver farm and were with people who loved them. They had no desire to leave.

Then one night, bushwhackers made a final raid on the Carver farm. As they came riding in, Moses grabbed James and yelled for Susan and Mary to hide. Moses and James hid in some bushes while Susan hid behind the barn. Pausing to wrap baby George in a blanket cost Mary enough time to find an escape. The bushwhackers took one look at Mary and congratulated themselves on finding a young healthy slave they could sell for a lot of money. Mary resisted them with all her might. In the end she was thrown on the back of a horse, still holding tightly to George, as she was forced to ride out with the bushwhackers.

Determined to get Mary and baby George back, Moses set out early the next morning and found one of his neighbors who was a Union scout. He offered him a race horse and prime land worth eleven hundred dollars to bring them back. The scout tracked them for five days until he could find no more trace of Mary, only hearing that she had been sold as a slave to farmers in another state. Then he heard about a scrawny little black baby who had been dropped off at a farmhouse. Giving up on finding Mary, he made it his mission to at least find baby George, who was barely alive when he was brought back to the Carvers. Susan was heartbroken over the loss of her friend Mary, but she was determined to nurse her very sick baby back to health.

James and George never saw their mother again after that night. Even though President Lincoln had already signed the Emancipation Proclamation, the order would not take effect until after the war ended.

Somehow, George survived as Susan carefully nursed him back to health. Even before he had reached his first birthday, he had already been forever separated from both his biological mother and father.

Uncle Moses and Aunt Susan became like George's surrogate parents. They raised George and his brother as their own children, impressing their strong work ethic on them. The Carvers prided themselves on being responsible and self-sufficient, able to make do with what they had, but that left little time for frivolous activities like reading or art or music. Although these were things that gave George much pleasure during his life, the Carvers did not encourage George or James to pursue them.

Although Susan and Moses were honest and hardworking, they seldom spent Sundays attending church. Almost everyone else in the community took time to meet in a local church on Sundays. One day George asked a neighborhood boy what they did at church. He replied that they just sang some songs, listened to a sermon, and closed their eyes and prayed, like talking to God. George decided he would try to talk with God. He began by introducing himself to God and found that praying was very natural for him. Prayer soon became a regular part of his life and became even more important after he was given a Bible of his own to read.

Review and Expand Your Knowledge

1. The land surrounding the Carver farm was described as law-less. What does that mean? In your opinion, would you like for your family to live in a lawless territory with no official laws or law enforcers?
2. Why did the bushwhackers choose to live in this region?
3. In 1861, what did some of the southern states choose to do about being part of the Union?
4. What did the bushwhackers plan to do with Mary after they kidnapped her?
5. What was the Emancipation Proclamation? Who signed it?
6. Which of the following did Moses Carver encourage the two boys to learn—music, art, reading, a strong work ethic?

CHAPTER 2

WAS IT A SIGN?

For as long as George could remember, he had a desire to learn to read and to learn more about things in nature. He believed that learning to read would open many doors for him. Later, in a letter to one of his professors, he wrote, "From a child I had an inordinate desire for knowledge, and especially music, painting, flowers, and the sciences, algebra being one of my favorite studies."

Before the Civil War, teaching a black slave to read and write, except for religious instructions, was discouraged and was even illegal in some states. Because of this practice, there were only a few black adults in the South who had been taught how to read and write. George knew Aunt Susan could read, but he also knew how the Carvers felt about wasting time when they could be doing something useful. With considerable reluctance, he finally asked Susan if she would teach him to read. To his surprise, she was happy to do this. Using an old spelling book, Susan taught George the letters and the sounds they made.

Soon he had memorized every word in the book. This was a great beginning for an education, but it only fueled George's desire to attend school and learn about new things beyond the farm.

As they became older, James worked in the fields with Uncle Moses, while George, who was frail

5

and often sick, learned from Aunt Susan to cook, clean, mend, garden, and prepare simple herbal medicines. George developed his own concoctions of pesticides, herbicides, and nutrients to improve soil. They became so useful to local farmers that he became known as "the little plant doctor."

The day came when Aunt Susan was finally able to tell George about the night he and his mother were kidnapped by a gang of raiders. He learned that Mary had held him tightly as she fought the outlaws with all her might. George had always held on to the hope that he would see his mother again, but as Aunt Susan spoke, he realized this would probably not happen. Although Susan and Moses were happy to see baby George a few nights after the raid, when Mary didn't come back years after the war was over and all the slaves had been freed, they knew she would have found her way back to her children and the farm if she could have done so.

After giving up on the dream that his mother would be coming back, George walked through the woods to grieve the loss of his parents who had been separated from him so young. He found peace in his secret places—places where the simple things he loved could be found. Things like plants and flowers and rocks and the sounds of animals in nature. He thought deeply about his life and wondered why he had not died as a very sick baby. Did his life have a purpose? How would he ever find it? A plan took root that day and grew into a burning desire as he became older. He needed books to read in order to answer the millions of questions he had about the things found in the woods. And he needed to go to school.

The Carvers taught George many practical things about everyday life. Uncle Moses showed George how to safely collect honey from beehives, keeping some and selling some. Aunt Susan taught him how to spin wool and flax into thread, make cloth, and sew the cloth into the clothes they wore. George learned to tan cowhides and make leather shoes and how to breed good hunting dogs. He didn't realize "book learning" and practical skills would blend perfectly to make his life successful in the future.

When George was about ten, an unexpected skill emerged from his creative mind. The Carver house was safe and comfortable, but it did not contain fancy or frivolous items. One day he looked into a neighbor's house and saw beautiful paintings hanging on a wall. Astonished and awed, he set about to find things in his woods that could be made into paints and brushes. Bark, stones, and old boards became his canvas as he invented paints from the colors in the soils and fashioned brushes from plants to create lovely artwork.

Around the same age, George had borrowed Uncle Moses's sharp hunting knife to help a neighbor finish making a set of crutches. He was enamored by the things a knife could produce, and he began to yearn for his own knife. Owning his own knife consumed his thoughts for weeks and was an even greater desire than learning to read. One night the desire for a pocketknife was almost overwhelming. When he finally fell asleep, he had a vivid dream in which he saw a half-eaten watermelon next to three stalks of

The farm house of Moses Carver (built in 1881), near the place where George Carver lived as a youth.

corn. On the dirt beside the watermelon sat a small two-bladed pock-etknife with a black leather handle. He woke up the next morning wondering if his dream might have had some meaning.

After finishing his chores the next morning, the dream still dom-inated his thoughts. He ran into the cornfield to look for three corn-stalks next to a half-eaten watermelon. He spotted them across the field exactly as they appeared in his dream. He cautiously approached the corn stalks, and his heart almost stopped when he saw a knife lying on the ground next to the watermelon! In the days that followed, George constantly reached into his pocket to touch the knife and to reassure himself it was still there. He began to believe this might be a sign that just as he received a knife he had often dreamed of owning, another dream to learn to read might also come true. Maybe God was telling him to pursue his dream, and He would take care of him.

Review and Expand Your Knowledge

1. What did Moses and Susan Carver do with the two children who had belonged to Mary?
2. Why were there only a few black people in the South who knew how to read and write after the Civil War?
3. Where did George go when he wanted to find a quiet place to pray and to think about answers to his questions?
4. What were some of the practical things about living on a farm that Moses and Susan taught George?
5. What led George to find a pocketknife that he really wanted? What did he think this might mean for his future?
6. At an early age George began to make plans for his future. What did his plan for his future include?

CHAPTER 3

CURIOSITY AND A THIRST FOR KNOWLEDGE

George's curiosity knew no limits, and he wanted to learn everything he could about plants, flowers, insects, animals, and even rocks. Aunt Susan taught him all she knew, but further education would need to come from school. However, George was not allowed to attend the nearby school in Diamond Grove, because it was for white children only.

Torn between his love for his family and a thirst for a real education, George left the farm at the age of eleven, not long after his brother, James, left the farm to find work in Arkansas. George left so he could attend an all-black school in the neighboring town of Neosho and begin his dream of going to school. Full of apprehension and fears, he walked the eight miles from the Carver farm to Neosho during the late afternoon to be ready for the first day of school with no plans for a place to sleep that night or where he might find food to eat. George and Susan expected him to return home after a few days since he had no food or money, but that's not what happened.

Mariah Watkins discovered him sitting on a pile of wood in her backyard. Politely explaining that he was waiting for school to open in the morning, George melted her heart. She invited him to come inside for a warm meal and to meet her husband, Andy. Before the night was over, George had been invited to stay with the childless couple and attend school in exchange for his help with Mariah's laundry business.

On the first day of school, George used his lunch period to help Mariah finish the waiting laundry. The clothes were sparkling clean and hung out to dry before he had to be back in school. He cheerfully did everything asked of him to the best of his ability. In return, the couple provided food and shelter, while Mariah taught her young helper wonderful lifelong lessons from the Bible.

His new family always attended the local African Methodist church on Sunday. George soon learned that the congregation expected four things from their members: to be kind to one another, work hard six days a week, go to church on Sunday, and to trust God.

Though the Carvers were hardworking, honest, and tried to be fair to everyone, they had seldom attended any church. Mariah's devout faith was lived out in her daily life according to the things taught in the Bible. She gave George a small, black leather-covered Bible and planted seeds of faith in him. He promised himself he would read it diligently for the rest of his life because he knew it contained the answers to his questions about God and life. Mariah also fueled his love for plants as she shared her extensive knowledge of medicinal herbs with young George.

Curiosity and a desire for more knowledge remained a burning desire in George's heart even after two years of attending Neosho school.

He was grateful for what he had learned from Mariah and from school, but at this point, he had a higher level of education than anyone in his class, including the teacher. His dream for a good education overcame his desire to stay in a safe place with the Watkinses. Still unsure that he could make a living on his own, he joined a wave of African Americans who were traveling west looking for better opportunities. He said a sad goodbye to Mariah and Andy as he left Neosho to make a journey to Kansas with a mule train. His knife was a reassurance that God would be with him, and his faithfully-read Bible gave him the courage to continue to pursue his dream.

George's first order of business in this new town was to find a place to live and secure an income. He found a friendly-looking house, walked around to the kitchen door, and knocked softly. A tall, blonde woman answered the door. George stammered that he was looking for a job, adding that he could do just about anything. When the woman asked if he could cook, George answered, "Yes, ma'am," although about the only thing he knew how to cook was beans with ham bones. He added that since most people had their own special way of cooking, he was willing for her to show him just how the family wanted their food cooked. After a few lessons, George was able to prepare fine meals that everyone liked. He happily moved into the servants' quarters since he had no other place to live and it came with plenty of food.

George had a plan and things went well for him for a few years. When his jobs allowed him to save enough money to buy books, he would quit working and go to school. Already eighteen, he needed a few more years to graduate. He was pleased that his plan was working until two terrifying events happened.

Review and Expand Your Knowledge

1. Why did George have to walk eight miles to attend school when there was a good school in Diamond Grove?
2. Arriving in Neosho with no food or money, how was George able to stay there for two years while he attended school?
3. What gift did Mariah give George that he used every day?
4. Where did Mariah and Andy go every Sunday?
5. How did George travel from Missouri to Kansas?
6. George moved from town to town several times while still in his teens. What did he do in each town where he lived?

CHAPTER 4

TERRIFYING OBSTACLES TO DREAMS

One day as he walked through town, two angry white men blocked his path. One of them shouted, "What are you doing with those books, boy?" Fear rose within as George replied that they were his schoolbooks.

The second man answered in a mocking voice, "That's ridiculous! A black boy pretending to read those books. No, boy, you stole them from a white child. Let me show you what happens to black boys who go around stealing white people's books."

Without warning, the first man hit George hard in the stomach with his fist, causing him to fall to the ground. He instinctively curled into a ball. The men took turns kicking him as they cursed and accused him of stealing. The pain was searing. People heard the commotion and saw what was happening, but no one ventured to try to stop the bullies. Even though almost no one approved of what the bullies were doing, they looked away, and walked on by in silence. Eventually, the two men tired of tormenting George and left, laughing about how they had taught the boy a lesson about stealing.

George crawled to the side of the road and propped up next to a tree. He looked around for his books and notes, but they were nowhere to be found. The money he had worked so hard to earn was wasted. He wondered if there was any use in pursuing his dream of an education anymore. He sat for a long time before he was able to get up and limp back to his room. When he felt better, he thought about his options but with less enthusiasm than before. Regardless of what he did next, he would need a job to pay for his room and food. Maybe he would try to go to school later if he could earn enough money to buy new books.

His new employer repaired horse harnesses. One day, he sent George to the Hall farm three miles out of town to deliver a newly repaired harness. As George walked past the jail that afternoon, he noticed a crowd was beginning to gather, talking to each other in intense, quiet voices. Returning home in the fading light, he found that the crowd had grown to about a thousand as more wagons headed into town. The crowd had changed from quiet and restrained to loud and angry. They began demanding action and broke into a chant: "Lynch him. Lynch him."

George quickly hid as best he could to be out of sight from the mob. He watched as men in white masks dragged a black man down the street. They threw a rope over a lamp-post and placed a noose around the man's neck. George couldn't

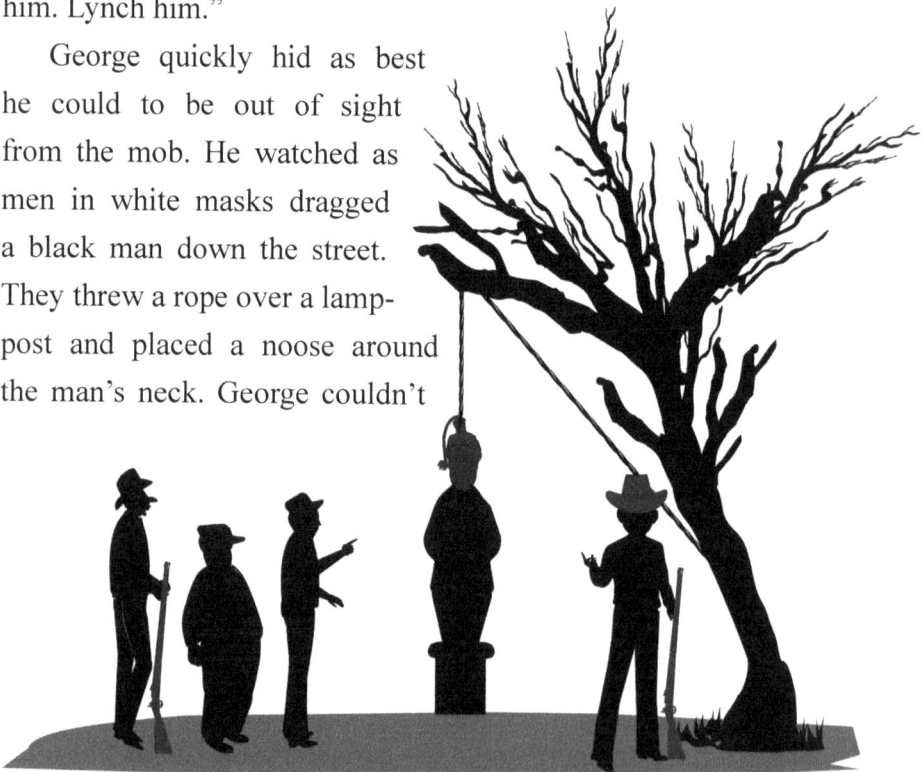

see what happened next, but a roar and cheering from the crowd told him the man had been hanged. He remained in the darkness for hours before he finally made his way back to his little cabin. The dreadful sounds and images of that night would be forever etched in his memory.

Though the black man might have done something awful, he was not given a chance for a fair trial, which is guaranteed by the Constitution. The mob ran roughshod over the Constitution that night. George wondered about the possibility that a white man had made up some story to cover his own misdeed. It wouldn't have been difficult to believe a white man over a black man. Would anyone believe him if he were accused of a crime he didn't commit?

Darkness still covered the town. Knowing he would never feel safe there again, George tied his few possessions into a kerchief, quietly slipped through the gate, and headed out on a northbound road for an unknown destination.

Review and Expand Your Knowledge

1. Why did the two bullies who hurt George assume he had stolen the books he was carrying from a white student?

2. Even if witnesses see someone committing a very bad crime, why is it still illegal for a mob of people to execute that person? What does the Constitution say about this?

3. What is the Constitution of the United States? What is the section of the Constitution known as the Bill of Rights? What are some of the main rights spelled out in the Bill of Rights of the U.S. Constitution to citizens?

4. Why is a court system designed to administer justice equally for everyone an important part of the United States?

5. If two grown white men were seen punching and kicking a young black person in the middle of any town in America today, which of the following would most likely happen?

    (a) Police would be called.

    (b) Someone would step between them and command them to stop.

    (c) Most everyone would ignore what was happening.

    (d) Most everyone would be too afraid to get involved.

6. In your opinion, which of the following finally established the principle of equal justice for all American citizens from 100 years ago?

    (a) New laws passed during the Civil Rights era making certain kinds of discrimination against minorities illegal.

    (b) A widespread rejection of the idea that the white race is the most superior race in the world.

    (c) Most citizens today support the idea that everyone deserves to be treated fairly and with equal justice.

7. In your opinion, would it be helpful to teach that white students are born with privileges that black students do not have, as some schools are choosing to do?

CHAPTER 5

VALLEYS AND MOUNTAINTOPS

George moved around and lived in several towns in Kansas during his teens and early twenties. He easily found work because he was always trustworthy, hardworking, and respectful. He knew how to set up a profitable laundry business and could organize a kitchen that turned out delicious meals. In each new town he implemented his plan to work until he had enough money for books and supplies along with daily living expenses and then to attend school as much as he could afford. While living in Minneapolis, Kansas, he could see he was getting close to his goal of graduation.

He tried to keep in touch with the Carvers and other friends he had met during his travels by writing them letters. One time he wrote letters to several people but got almost no replies. When he asked the postmaster if there was a problem, he discovered there was another person named George Carver living in the same town. The postmaster thought the other George Carver was probably getting all the mail because no one expected a black man to get a lot of mail. George was advised to use his second initial to distinguish between the two men. Since slaves typically had no birth certificate and no official middle name, he just randomly chose "W," making his name George W. Carver. One of his friends asked if the W stood for Washington. The name struck a chord with George, who thought Washington was an impressive middle name to have. From then on, he signed his name as George Washington Carver.

One day he received a letter from Moses Carver containing the bad news that his brother, James, had died from smallpox. The letter left him with a deep loneliness when he realized he had no living relatives, although his guardians, Moses and Susan Carver, and a host of other friends cared for and loved him greatly. He dealt with his sadness by affirming his trust in God, reading his Bible, and playing the accordion out in the woods. Getting up each morning before dawn to walk in the woods and fields while he meditated on things he read in the Bible and things in nature that inspired him, was a daily practice he continued as often as he could.

Then in late 1883, he finally graduated from high school. His teachers recognized his unique skills and abilities and encouraged him to apply to attend college. He set his sights on Highland College, a small Presbyterian school in Kansas. After mailing the application, he waited anxiously to see if he would be accepted. The day came when he received the letter saying he could attend. With great excitement, he arranged to sell his laundry business

James Carver

and planned a trip to visit Uncle Moses and Aunt Susan and tell them the good news. At the top of his agenda was a visit to James's grave.

The visit with Moses and Susan went well. His loving, adoptive guardians told George how proud they were of his achievements and decisions and sent him out with their blessings. By the time August arrived, George had all the money he would need to pay for a year of college.

A few weeks later, he boarded a train for Highland, Kansas. Excitement built as he located the college and began walking toward the large central building. The first person he met inside was an elderly woman who viewed George suspiciously. She finally asked, "What do you want, boy?" To which he replied that he had come to enroll. "Wait here," she said in a condescending tone.

Finally, a tall man strode toward George and said in a cold voice, "You did not tell me you were a Negro. This college does not accept Negroes."

Speechless and more than a little disappointed, George knew there was no point in arguing about the unfairness of the situation. Without a word, George turned and walked away. He had fallen from hope-filled excitement to hopeless despair in a matter of seconds. He kept walking until he found a church where he heard singing inside. He sat down outside to ponder this sudden roadblock to his dreams. When the service was over, several women found a dejected young man sitting under the trees. They greeted him warmly and asked what brought him to Highland. If George had not been so depressed, he probably would have left the town that had crushed his dreams rather than tell these white women his life story.

Surprisingly, they sympathized, and one lady even offered him a job. George was grateful for a job, but after working as a domestic helper for the next year, he realized he might be doing this for the rest of his life unless he took another route. One of his new church friends encouraged him to take advantage of the Homestead Act of 1862 that allowed settlers to file a provisional claim for 160 acres of land, work the land for five years, and then file a claim to own the land. George thought about this for the next few days and decided to start the process.

Confident that he could be successful as a farmer, George found employment as a farmhand helping a fellow settler build a barn and hen-house and later a house in which to live. Taking a liking to his hardworking young helper, George's mentor showed him how to build a house from sod that would survive the brutal winter weather that was coming and then file a claim for the land. George was surprised by the fierce winter storms he experienced, but by spring, George was living in his own sod house, had filed his land claim, and had bought essential household items from neighbors.

Then a conversation with another church friend renewed his dream of attending college. His thoughts constantly switched back and forth between being a farmer with his own land and trying one more time to finish college.

As he had done since his teens, George continued his practice of starting his day by taking early morning walks or spending quiet times with God outside as much as possible. He missed his beloved woods, but even though surrounded by flat, bleak prairie, he continued to set aside time to think and meditate and commune with God, whom he called Mr. Creator. He was excited about his new house and the prospects of acquiring 160 acres of land, but gradually the old dream to attend college revived, overshadowing the possibility of a deed to his own land. He spent a lot of time during his morning meditations praying about what he should do. He almost never made hasty decisions or changed the direction of his life without his early morning conversations with God. Morning after morning, an inner prompting urged him to continue the dream to attend college. He finally made the difficult decision to relinquish his land claim and head north. As with most of his other departures, leaving good friends was sad and hard to do. But he knew in his heart this was the right decision.

Review and Expand Your Knowledge

1. How did George Carver get his middle name?
2. How did George respond to learning that his brother James had died from smallpox?
3. Why was George not allowed to attend Highland College?
4. Where did George find encouragement after he was denied attendance at Highland College?
5. What was the Kansas Homestead Act of 1862?
6. Why did George choose to give up his land claim and instead try one more time to go to college?

CHAPTER 6

THE ENCOURAGERS AND GOD'S GUIDANCE

In the spring of 1888, George arrived in Winterset, Iowa, and looked around for a place to stay for the night. The St. Nicholas Hotel was a possibility but more than he could afford. About to walk away, he noticed a sign advertising for a cook to start immediately. Half an hour after inquiring about the job, he was employed at the hotel and had a place to live. By Sunday, he had found the welcoming, supportive Winterset Methodist Church where he met a white couple, John and Helen Milholland. As a deep friendship developed between them, they recognized George's desire and ability to pursue a higher education. They encouraged him to apply to attend Simpson College, a Methodist school with no restrictions based on race. Again, he applied to attend a college. He was accepted and scheduled an interview with Mr. Holmes, the president of the school.

As George walked to the campus of Simpson College, old fears rose in his mind. When he arrived, he found Mr. Holmes helpful and welcoming, although a problem remained.

All of the male students boarded in homes throughout the town. Mr. Holmes doubted the local town folk would want a black man living in any of their homes. He then made George an offer to allow him to live in a run-down shack at the back of the college campus. To Mr. Holmes's surprise, George enthusiastically accepted.

All of his life, George had made the most of his opportunities and his supplies. Later the same day, he went shopping and purchased cleaning supplies, two washtubs, a washboard, matches, and some starch. With his last fifteen cents, he bought some cornmeal, and a little beef fat for his next few meals. Before the week was over, he was happily scrubbing his new shack, had set up a small laundry business, and was cooking his simple meals. Wooden packing crates served as furniture. He had finally made it to college, and even his meager circumstances could not discourage him.

Before long, he had a thriving laundry business. Many students would come by and chat as he finished their laundry, impressed with his intelligence and knowledge and determination. As others became aware that George was living out of crates, mysterious gifts began to appear. Throughout the year, tables, chairs, beds, warm blankets, concert tickets, dollar bills, and various other items just materialized in the little shack—always anonymous so he could not refuse them.

George began his college studies pursuing art and piano, often painting beautiful, detailed pictures of the flowers and plants he collected. One of his professors recognized his vast knowledge of plants and his deep interest in learning more about them. She invited George to sit down with her and have an honest, realistic conversation about his future as an artist. George admitted that although art fulfilled a place within his heart it was not his first love. His professor noted that as a black man he was not likely to have a successful career as an artist anyway. Instead, she encouraged him to make art a hobby and pursue an education in botany at the Iowa State Agricultural School. After thinking hard about what she said and seeking God's advice during his early morning meditations, he took his professor's advice and applied to attend yet another college. George was hopeful that all would go well, but he also accepted the risk of being rejected in a new school.

Review and Expand Your Knowledge

1. Whenever George arrived at a new town, he followed his plan to look for a job and a place to stay. He would then save his money to buy books and pay for other college expenses. What kind of job did he find at Winterset, Iowa?
2. Where did he go on his first Sunday in town?
3. Where did new friends at church recommend that he apply to attend college?
4. How did the interview go between George and Mr. Holmes? Why were George's meager living conditions satisfactory to him?
5. What business did he start to pay for his college expenses?
6. What kind of courses did George take at this college?
7. What changes did one of his professors advise him to make that involved moving again?

CHAPTER 7

NEUTRALIZING RACIAL PREJUDICES

In spite of his misgivings, George easily gained admission. Leaving Simpson where he experienced genuine friendships and where his fellow students and professors had showered him with kindness was especially hard. Nevertheless, logic and a familiar inner guidance told him this was the way he should go.

After beginning classes at Iowa State, George had serious doubts that he had made the right decision. He was the only black student on campus. His room consisted of an old storeroom that had to be cleaned out. No one wanted to sit next to him, and he was aware at times of jeering remarks aimed at him. The dining room manager demanded that George eat in the basement with the black servants. He could do nothing except endure the humiliation and pray things would get better.

He finally wrote a letter to his friend Mrs. Liston at Winterset about the situation, expecting some words of wisdom from her in a return letter. However, the next day, Mrs. Liston showed up on campus looking for George. She walked arm in arm with him as George showed her around, fully aware that students were whispering to one another about why a white woman would be so friendly to George.

When it came time for lunch, George suggested that Mrs. Liston might want to eat dinner with the white staff and students. She insisted on eating in the basement with George and the black workers. When one of the faculty members came downstairs to invite Mrs. Liston to eat upstairs, her reply shocked even George. "My good Sir, George Carver was one of the best students at Simpson College before he came here. My husband and I consider him like a son. Wherever you decide he should eat is good enough for me!" By early afternoon, Mrs. Liston had left, leaving George with mounting fears about how her visit would affect his treatment by students and faculty in the future.

George knew he had not been accepted, much less liked, by anyone at the college so far. What if they all hated him now? He could be met with more disrespect than ever and even by hostile actions. He spent the night trying to imagine how he would react to these possibilities.

The next morning, George followed his normal routine, but dreading the day ahead. He was unprepared for the treatment he was about to experience. Several students smiled and spoke to him, and one student invited him to eat lunch upstairs with him. George realized that students and faculty were not all full of racial hatred. They just had

Statue of George Washington Carver at Iowa State University

preconceived ideas that all blacks were not intelligent enough to do college work and were only capable of being servants. Mrs. Liston had exposed the racial prejudice many students and faculty had been practicing without even realizing what they were doing. Once they recognized their biases, they were ashamed of their behaviors and began to set them aside and look at George as an equal.

Her visit was a positive turning point for George. As other students and faculty got to know him better, he became accepted for his shining character traits, skills, and knowledge. Before the year ended, he was inundated with invitations to join an assortment of campus organizations.

One of the organizations he found most interesting as a student was the Young Men's Christian Association (YMCA). The purpose of this organization was to look for ways to encourage racial equality in the South. Not an active member at the time, he encountered the group again decades later when he was asked to be the keynote speaker for a YMCA summer camp.

The mostly white students who later attended summer camp received his speech enthusiastically. They cheered as he talked about the need for equal rights for all. This was a positive sign of changes in race relations, but support for racial equality was found mostly among young people and not as much among the older generation.

George was thankful that race was not a barrier to knowing and hearing from his Creator. He thanked God for His guidance in helping him make the hard decision to transfer to Iowa State and then persist through a difficult time without giving up. During the next three years, he earned the reputation among his professors as one of their most brilliant students. He developed deep personal friendships with two of his white professors, Henry Wallace and James Wilson, who guided him as he developed his unique abilities and interests. Their paths would cross again in the future.

Uncle Moses and Aunt Susan instilled strong work ethics in George when he was young. He was never too proud to work to pay for his room, meals, and classes. Doing odd jobs for faculty, working as a janitor, tutoring other students, and doing laundry for others funded his education. The football coach even paid him as a masseur when players realized George had a knack for massaging and relaxing sore cramped muscles.

During George's time at Iowa State, he learned new farming methods and scientific techniques that would increase production and the quality of plants grown by the farmers he would soon meet. He learned ways to keep the soil healthy and fertile. He was trained in how to analyze soil to discover important substances such as nitrogen, potassium, phosphorus, and other nutrients that had been depleted from the soil, and he learned practical ways to replenish these substances. He became an expert in crossbreeding plants and was able to produce improved varieties of several plants.

Graduating from the agricultural school in 1894, George became the first African American to earn a Bachelor of Science degree from Iowa State College of Agriculture and Mechanical Arts. He grieved when his beloved guardian, Aunt Susan, died before she had a chance to see him finally achieve his long struggle to finish college.

After graduation, he was faced with the decision whether or not to continue his education and work on a Master's of Agriculture degree. However, he didn't have to think long about this. His research on fungal infections in plants had so impressed his professors that the school invited him to stay and earn another degree and even hired him to join the faculty as a professor. His new position enabled him to earn a living by teaching instead of doing laundry. He soon discovered that he was a natural teacher who had a gift of simplifying complex ideas.

During this time, George had the privilege of working at the Iowa State Experimental Station with L. H. Pammell and co-authored two books with him about treating and preventing fungal diseases in cherry trees. Working with Dr. Pammell, who was a nationally known mycologist (fungal scientist), George became an expert in his own right in identifying and treating plant diseases. Nationally, his expertise in mycology was second only to that of Dr. Pammell. His years at Iowa State equipped him to pass on valuable farming information and methods to thousands of future students, as well as many poor southern farmers.

Review and Expand Your Knowledge

1. George said goodbye to many friends at Winterset College who appreciated his character and intelligence. How was George accepted at Iowa State College in the beginning?

2. When Mrs. Liston learned that George was being disrespected at the new college, she paid a visit to him. How did her visit help him find acceptance among the other students and faculty?

3. Did George become bitter and angry with the ways in which he was first treated by other students and faculty at Iowa State?

4. Did George believe all white people had hateful, racial attitudes toward black people?

5. What kinds of jobs did George take to pay for his college expenses?

6. What new kinds of farming methods and scientific techniques did George learn that would enable him to help farmers in the future?

7. What kind of jobs did George have while he was finishing his Master's degree?

8. What sad news did he receive before he finished college?

9. What unique achievement did George accomplish at the university in 1894?

CHAPTER 8

A NEW CHAPTER IN LIFE

George completed his master's degree in 1896 at the age of thirty-two. His degree was not the end of his dream. Instead, it opened the door to what he was uniquely equipped to do. His early morning meditations and the time spent communing with God had focused his goals on spending the rest of his life serving and helping farmers be more successful in growing crops. He was especially interested in working with poor black farmers. His objectives eventually expanded to help struggling farmers of all races, and even some wealthy corporations.

After graduating, George was invited to retain his position as an instructor at Iowa State College. He also received several offers for employment that included generous salaries in places in which he would be greatly respected. However, the one he found most interesting did not come with either benefit. It came from Booker T. Washington to head the agriculture department at Tuskegee Institute in Alabama. This position would allow him to teach poor southerners to use better methods for planting, growing, and harvesting crops, while still doing agricultural research. Washington's letter stirred George's heart.

Tuskegee Institute seeks to provide education—a means for survival to those who attend. Our students are poor, often starving.

> *They travel miles of torn roads, across years of pov-*
> *erty. We teach them to read and write, but words cannot*
> *fill stomachs. They need to learn how to plant and harvest*
> *crops.*
>
> *I cannot offer you money, position, or fame. The first*
> *two you have. The last, from the place you now occupy, you*
> *will no doubt achieve. These things I now ask you to give*
> *up. I offer you in their place—work—hard, hard work—the*
> *challenge of bringing people from degradation, poverty,*
> *and waste to full manhood.*

George could have continued to work at Iowa State as a respected faculty member and conduct the kind of research he loved with a well-financed and organized greenhouse. Yet Mr. Washington's letter made a connection with the deepest part of George's being. He paced the floor as he considered his options, praying for an answer. Many of his friends thought he would be crazy to give up a perfect job where he was highly respected to move to a place like Alabama where he might not be treated fairly because of his race. He knew the problems for a black man living under Jim Crow laws that supposedly provided "separate but equal" facilities and opportunities, but in reality, limited many of their civil rights.

Nevertheless, his reply to Mr. Washington's letter left the door open to accept a job with Tuskegee. George wrote, "Of course it has always been the one great ideal of my life to be of the greatest good to the greatest number of 'my people' possible. To this end I have been preparing myself for these many years; feeling as I do that this line of education is the key to unlock the golden door of freedom for our people."

Nine days later, another letter from Tuskegee arrived. Mr. Washington wrote that he wanted an all-black faculty that would be good role models for the students. He said he considered George the only black man in the United States, possibly the whole world, with the qualifications to run the new agricultural department. His offer included a salary of "$1,000 a year, a room in the student dorm, and food at the cafeteria."

After days of seeking God's guidance, George made his decision. He was going to Tuskegee. The deep friendships George had developed during his years at Iowa State made leaving especially hard for everyone. He left his dear friends at Iowa State College with their gifts of a new tailor-made suit and a microscope, both of which served George for decades. Once he made his decision, he confidently wrote a letter of acceptance to Tuskegee, saying:

> *I am looking forward to a very busy, pleasant, and profitable time at your college and shall be glad to cooperate with you in doing all I can through Christ who strengthens me to better the conditions of our people. Some months ago, I ready our stirring address delivered at Chicago, and I said amen to all you said.*
>
> *Furthermore, you have the correct solution to the "race problem."*

Booker T. Washington believed southern farmers, black and white, who were struggling to make a living should look for ways to work together to help one another. Both men agreed that civil rights for blacks would be hastened as black farmers proved their skills and abilities alongside white farmers. Neither man would live to see full civil rights for all and the changes Dr. Martin Luther King would bring about in the future. For almost half a century, George played an important role in helping southern farmers, black and white, improve their living conditions. And along the way, he helped to pave the road for civil rights for all.

However, successful farming was not what George saw from the train windows during his ride to Tuskegee. The scenes revealed the enormity of the problems George would deal with for the next several decades. He wrote of seeing "acres of nothing but cotton." He was horrified that "everything looked hungry: the land, the cotton, the cattle, and the people." Decades of planting nothing but cotton on the same fields year after year had sapped the land of its nutrients. He arrived at Tuskegee with a mixture of excitement about the possibilities for teaching future farmers how to improve the productivity of their farm, along with an awareness of the difficulties of achieving such a lofty goal.

George Washington Carver (front row, center) poses with fellow staff members at the Tuskegee Institute (now known as Tuskegee University)

Although the campus was clean and free from clutter, it was even worse than the farms, with most of it in a swampland. Classes met in shacks, and there was no laboratory or greenhouse. The time to build these facilities would have to be found in between a busy schedule of teaching classes and performing other duties.

George wasted no time in getting to work at his new position. He would work at the Institute for the rest of his life with great satisfaction. In his future were lucrative offers to work directly for wealthy businesses, which he turned down in order to fulfill his visions of helping downtrodden men and women find success and prosperity as farmers.

Review and Expand Your Knowledge

1. When George finished his master's degree in 1896, he was invited to retain his job as an instructor at Iowa State College. He also received several other offers for employment, all of which he declined. What kind of work did George envision for the rest of his life?

2. Who sent him the offer for a job that he was most interested in accepting?

3. Think about why George chose to give up a good-paying job at Iowa State College where he was highly respected and could do the kind of research he loved in order to move to Tuskegee Institute, one of the poorest schools in America. In your opinion, why do you think George chose Tuskegee?

4. What salary and benefits did Booker T. Washington offer George to move to Tuskegee as the Dean of the Agriculture

Department? What three things did he ask George to give up to work with him?

5. How did the Jim Crow laws in the Deep South limit the civil rights of black citizens during this time?

6. For a few years, George Washington Carver and Martin Luther King Jr. were contemporaries. Without a doubt Dr. King heard of the work of George Washington Carver, but he never had a chance to meet him. Both men deeply desired a time when there would be full civil rights for all citizens. In your opinion, do you think George would have approved of Dr. King's method of nonviolent protests?

7. The only source of income for many southern farmers was cotton crops. Why did George advise farmers to practice crop rotation or change the kind of crops they planted every year?

8. What conditions did George find on the school campus when he arrived at Tuskegee?

9. What conditions did he find on most of the farms of the poor farmers in the South? Did George know practical ways to make farming more prosperous?

CHAPTER 9

GROWTH FROM A MEAGER BEGINNING

The Tuskegee Normal and Industrial Institute was different in many ways from other schools of higher learning. Everyone who applied was accepted, even though many applicants could barely read and write. With little equipment or money to start the agricultural department and only thirteen students signed up for agricultural studies, George realized he would have to take steps to get more students enrolled in his classes.

His next goal was to locate a few supplies for the science laboratory. The only piece of equipment in the whole department was a microscope, which had been given as a gift to George from Iowa State College when he left for Tuskegee. In one of his first classes, George took his thirteen students to a trash pile in the back of the school and assigned them to search through the trash and collect anything that looked like it could be useful to the lab. Bottles became test tubes. Holes drilled in the bottom of a worn-out pot turned it into a sieve. Spokes from an old wheel were fashioned into a tripod.

The overall need for the agriculture department was to find money to finance a laboratory building and equipment. Tuskegee did receive some government funds, although only about a tenth of what was awarded to white colleges. Nevertheless, the funds were allocated carefully, and the agriculture department eventually received enough money to build a much-needed facility. Within a few years, a new modern two-story structure with an herbarium, labs, and lecture rooms accommodated twenty-seven students enrolled in the agriculture department.

Then George received an invitation from James Wilson, one of his old professors who was now the U.S. Secretary of Agriculture, to visit him in Washington, D.C. George accepted the invitation and had a chance to tell Secretary Wilson all about Tuskegee Institute. During his visit to Washington, George was also able to secure funds from the president of the Smithsonian Institution for more research equipment.

George returned the invitation and asked Wilson to visit Tuskegee and dedicate the new Slater Armstrong Memorial Agricultural Building in the fall. Wilson was happy to do this and to promote Tuskegee at the federal level. An enthusiastic student body and faculty turned out to prepare for the visit of Secretary Wilson. This was the first time ever that someone from the executive branch of the federal government was coming to visit Tuskegee. Trees and bushes were trimmed, floors were scrubbed, a choir practiced daily to welcome him, the grass was cut and raked, and even a huge fireworks display was ordered. Secretary Wilson's visit did much to improve the prestige of the college and aid in fundraising.

George was passionate about imparting useful knowledge to his students, but he was equally passionate about helping poor struggling farmers in the South who desperately needed to know how they could improve their farming methods. His students were often taught in unconventional ways. He added life lessons, farming techniques, and anything relevant to botany to his lessons. His Tuskegee students loved him. Why couldn't he reach the poor farmers in the same way?

Eventually, a few local farmers began to come to Tuskegee to ask George questions about farm-related problems. George decided to invite local farmers along with their wives once a month to attend a Farmer's Institute. They could ask their questions, and George would have an opportunity to show them new farming practices. Furthermore, they could learn from each other's successes and failures.

The farmers began to bring soil samples for George to analyze. If some of the essential elements were missing, he would show them how to put those elements back into the soil using inexpensive methods. He explained that cotton plants took much of the nitrogen and other important nutrients from the soil. An easy way to replace the nitrogen was to plant legumes, such as peanut plants, cow peas, and soybeans in the same soil and later plow the roots under the soil. He showed them that marigolds planted beside cantaloupes helped prevent pests. The experimental farm run by the students allowed him to demonstrate many useful techniques for improving the productivity of their land. One of his favorite pieces of advice was a suggestion for farmers to save a nickel a day to buy more land. By saving for a year, they would have $18.25, in which they could buy a few acres each year.

The Farmer's Institute, begun in 1898, was very helpful to the local farmers, but hundreds of other poor black and white farmers were unable to make a trip to Tuskegee. George needed a way to go out to these farms and share his knowledge about making farming more prosperous.

His idea took on life when he stepped out in faith, loaded up a wagon pulled by a mule, and drove into the Alabama countryside with enticing demonstrations of good farming. He trusted that someone would give him a place to stay for the night in exchange for valuable information. Whenever he came to a farmhouse, he stopped and talked. He showed the wives the jars of jams, jellies, and preserved fruits students had made at the Institute in Tuskegee. He encouraged the men to raise enough food to feed their families all winter and regularly save money to buy more land. He shared the lessons learned from Moses and Susan Carver about making some of their own tools and being self-sufficient. He encouraged farmers to plant sweet potatoes and other vegetables in addition to cotton. He researched new ways to use sweet potatoes and wrote a bulletin about how to grow and use them.

Booker T. Washington was determined to find enough funds to act on George's idea. When he learned of a New York banker named Morris Jesup who wanted to do something to help poor black farmers in the South, he quickly contacted him and secured an agreement to help fund the project. The Jesup Agricultural Wagon was equipped with updated information and farming equipment, and traveled through miles of farming communities, going to rural farms where farmers were unable to travel to Tuskegee. Eventually more than two

thousand farmers a month heard about new, tested methods of farming through the Jesup Wagons. Originally traveling by mules, later models went out in pickups and became known as the Tuskegee Institute Movable School. The successful farming techniques program expanded until it included water conservation and nutrition instruction, with modest funding by the U.S. Department of Agriculture and the Alabama Cooperative Extension System.

Before long, people from all over the world were taking notice of George. Well-known inventors, businessmen, and politicians began to make their way to Tuskegee to seek his advice or to offer their help. Around 1900, Andrew Carnegie visited the school in person to see what George was doing and donated a much-needed library to the college. George welcomed the president of Harvard University and the Secretary of War, William Taft, to Tuskegee. A delegation from Africa arrived to learn how to produce milk from soybeans, since they had difficulty keeping cows alive during hot drought conditions.

All of this gave Tuskegee much credibility and helped improve the school, but George wanted to make a difference for the farmers, both black and white. The terrible conditions under which poor farmers lived cried out for someone to help. George had answers, but they involved abandoning or reducing cotton crops. How could he get them to listen and bravely make changes when local farmers had only planted and harvested cotton for decades?

Review and Expand Your Knowledge

1. List some of the problems George identified at Tuskegee as he first began to work there.

2. What were some of the ways in which George was able to make improvements in the school?

3. In your opinion, why were poor farmers reluctant to stop planting cotton as their only cash crop?

4. What was the Farmer's Institute? What were some of the helpful things it accomplished?

5. What was George's idea to take this information to farmers who did not live near enough to attend the Farmer's Institute?

6. How did his idea of going out to farmers in remote areas greatly expand and get funded?

7. What did Andrew Carnegie do to improve the Institute?

8. What other influential people visited Tuskegee to see the work George was doing?

9. Why did a delegate from Africa come to Tuskegee to visit George?

CHAPTER 10

A DANGEROUS BRUSH WITH RACIAL HATRED

Small rural schools that taught black children to read and write, do basic math, and work the land sprang up across the South during the early 1900s. Many of the teachers were George's former Tuskegee students who had caught his enthusiasm for applying new farming methods to local farms and learning to read. George visited as many of these little schools as possible, offering advice to current and future farmers and encouraging the teachers.

One of these schools was located in a small community several miles outside of Montgomery, Alabama. The racially segregated town did not allow black people to live within the city limits. Nevertheless, the schools and their teachers were respected and well-liked by most people in the area. A famous photographer from the North, Frances Johnston, heard about the schools. She made plans to visit Tuskegee and then go with George to see as many of them as possible. She wanted to document and photograph the improvements being made in the education of black children.

Her first day almost resulted in a tragedy. George was dropped off at the cabin of Nelson, one of the male teachers at the school. Nelson then offered to take Frances in the school's buggy to a hotel where she could stay during her visit. He planned to return to the cabin and meet with George.

When they arrived at the hotel and Nelson was about to unload her luggage, three angry white men on horseback descended on them. They were outraged to find a black man and a white woman together about to enter a hotel. Frances and Nelson were terrified as the men began yelling and shooting. Knowing that Frances was not their target, Nelson told her to run as he jumped back in the buggy and left at full speed to escape the bullets. Frances made her way back to the cabin, not knowing what might have happened to Nelson.

George understood the pockets of dangerous racial prejudices that persisted in the South. He knew the situation would escalate if Frances was found alone with George in Nelson's cabin. He helped her locate a safe place near the train station to hide the rest of the night until she could catch the early morning train to get back to Tuskegee. Then he would try to find another train to get himself out of the area and back home.

George walked for hours throughout the night looking for railroad tracks that would lead him to a different train station. He hid whenever he heard the vigilante group who was still looking for the black man seen with a white woman at a hotel. In the early morning light, he found what he was looking for and waited out of sight for the next train. An hour later, a train stopped. Weary and relieved, he climbed aboard and headed back to Tuskegee to see if Frances and Nelson were safe.

Frances greeted him with great relief and let him know that Nelson was fine too. Frances was furious that this was happening in the United States, and wanted to press charges, but Booker T. Washington thought that such legal action would do more harm than good. She did, however, make pictures depicting injustices in the South and pictures of fledgling black schools, as well as photographs showing life at Tuskegee. Some of those photos still survive in museums today.

The racial hatred revealed in the previous day's situation was not something George could easily forget. He realistically understood that in many communities throughout the South there were pockets of white people who considered it their duty to maintain tight control over black people and also bully any reluctant local whites into agreeing with them. They were willing to commit acts of violence, if necessary, in order to maintain a white supremacy philosophy over everyone in their community. Their loud forceful language and actions discouraged both white and black people from publicly protesting their controlling methods. The worse outcome of the racial encounter was that the black children in the school where Nelson taught were denied the opportunity to attend a good school and learn to read and write and study math, because the school had to close for months.

Disappointed by the outcome of the previous day, George gave in to rare feelings of self-pity. He wondered if white supremacy and racial prejudices would ever end, as he remembered that he had given up lucrative career opportunities that came with great respect, in order to use his knowledge to help the poor Southern farmers, black and white.

George spent a long time with Mr. Creator the next morning. As he paused to consider if divine guidance had led him to Tuskegee, he recognized a trusted inner voice assuring him that he had indeed been guided to make the many decisions that had led him to Tuskegee. And yes, his life had a greater purpose than that of his own personal ambitions.

George was determined not to let the events of the previous day dissuade him. He continued on the same path as before and reminded himself that for every hate-filled white person, there were many more white people who looked past his skin color and appreciated his character, knowledge, and abilities. He refused to judge all white people as hate-filled and prejudiced.

Nor did he turn down opportunities to speak to white audiences. Once he had been invited to speak at a farmers' conference that was attended by Dr. Walter Hill, Chancellor of the University of Georgia, and a white man. Hill's enthusiastic endorsement of his lecture was published in national newspapers and resulted in many more invitations to speak to a variety of groups, which he accepted as often as possible.

After a few days, he was soon back doing the things he loved— teaching, doing research, and even giving speeches. His days were filled with ways to share his knowledge about plants with future farmers. He found it natural and easy to teach his young students how to make a good living by farming, as well as how to live by the words of the Bible. His methods for growing better crops were being adopted by a number of farmers. He could feel God's approval for his persistence in holding on to his dreams.

Review and Expand Your Knowledge

1. How important were the schools that were organized and built for black children? In the early 1900s, were any of the schools in the South integrated where both white and black children attended together?

2. What was the purpose of Frances Johnson's trip to south Alabama?

3. Why did some angry white men threaten to kill the young black man who was helping Frances Johnson into the hotel where she had planned to stay?

4. Were George, Frances, and Nelson all able to get back to Tuskegee without any injuries?

5. Several years ago, there was a big news story about an American magazine publisher who had to go into hiding to protect his life. The publisher had printed a political caricature of the leader of an extreme religious sect. Some members of this sect felt it was their duty to kill the publisher whom they believed had insulted their leader. In your opinion, do you think murder or violence is ever justified when a tradition or person is insulted?

6. What is meant by an idea known as white supremacy? In your opinion, do you think most Americans today have rejected white supremacy?

7. At one time in the past, white supremacist bullies tried to maintain a tight control over what black people could do. What kind of tactics did they use?

8. Did the three men who rode up shooting their guns at Nelson cause George to change his mind about how he viewed all white people in general? How did he feel about speaking to white audiences?

9. Did George believe God intended for him to only use his knowledge to help black people?

CHAPTER 11

NEW ROLES AS SCIENTIFIC RESEARCHER, SPEAKER, AND ADVISOR

As George became widely known for his work at Tuskegee, he was asked to take on more and more responsibilities. His speeches and research projects were reported in numerous newspapers and magazines. He received many invitations to speak about his discoveries which came from simple plants. He served as an adviser to many well-known businessmen and government agencies.

Still, no matter how pressed he was for time, George continued his early morning meditations in the woods. He told a student, "Nothing is more beautiful than the loveliness of the woods before sunrise. At no other time have I so sharp an understanding of what God means to do with me as in these hours of dawn. When other folk are still asleep, I hear God best and learn His plan."

Continuing to experiment on ways to use peanuts, George also added soybeans to his research in 1904. He anticipated that soybeans would become a big cash crop in the South, because so many healthy, low-cost products could be made from them. This was an accurate prediction, but it wasn't fulfilled until many years later.

He also promoted the use of sweet potato flour to make bread. The idea of making sweet potato flour won him a trip to Washington, D.C., to speak to a Department of Agriculture committee. Skeptics at the meeting doubted the product would be practical until George took the entire committee to a local bakery and made loaves of aromatic, fresh-baked breads from sweet potato flour and served everyone samples.

The publicity of such widely-reported meetings resulted in George receiving several good-paying job offers, which he always declined. His heart's desire to help the poor farmers in the South be successful farmers and to share God's love for all never changed. Although briefly tempted, George knew his work was at Tuskegee.

By now George had become a popular speaker, and he continued to speak before influential committees and government officials as much as he could find the time to travel. Sometimes first impressions of George were not good. The first view of George seen by most audiences was an old black man wearing his trademark thirty-year old suit with an interesting weed in the lapel, and a brightly colored tie. Dragging his well-worn suitcase behind him, they often wondered if he had anything significant to tell them. But, more often than not, he left these meetings with the audience's great respect.

His original job at Tuskegee was as a teacher. Over the years, his responsibilities expanded to become a scientific researcher, a speaker, and an advisor to businessmen and government committees.

Although he had solutions to help farmers improve their crops, getting old traditional farmers to accept advice from a young, college-educated black man was a different story. Then a devastating problem entered the picture that caused many people to seek George for his advice.

Review and Expand Your Knowledge

1. Why did George decline several well-paying jobs with good equipment and labs?
2. Why did George start his day before sunrise in the woods?
3. Why did George begin to spend more of his time as a scientific researcher and speaker in addition to his teaching duties at Tuskegee?
4. What new crop did George correctly predict would become a valuable cash crop in the future?
5. How did George convince a Department of Agriculture committee in Washington, D.C. that sweet potato flour was a good, inexpensive source of food?

CHAPTER 12

A CONVERSATION
WITH MR. CREATOR

One of George's greatest frustrations was trying to persuade farmers they needed to include other crops than cotton. Crop rotation was an idea he often proposed. After talking with George, they would agree this was a good idea, but then they would go out and plant more cotton. During the early 1900s, George received an urgent letter from the U.S. Department of Agriculture, warning that boll weevils were already found in cotton crops in Mexico. They were headed toward Alabama. There was no feasible way to stop the oncoming army of insects that devoured undeveloped cotton bolls and ruined entire cotton crops.

George did everything in his power to warn farmers of the coming disaster. Some listened, but many did not. In 1904, the little weevils made their way into Alabama cotton fields. The farmers who had taken George's advice were able to harvest crops of cow-peas, sweet potatoes, and peanuts, giving them food to eat during the winter and cash for some of the crops. Not losing everything on a cotton crop was a good thing, but now farmers faced a new problem. Many farmers had a surplus of peanuts and few merchants to buy them. How were the farmers going to earn a livelihood from their new crops?

Distraught peanut farmers were desperate for answers. At least they weren't starving, but their crops weren't bringing enough profit to continue farming. George went out day after day for his pre-dawn meditations and prayed to God for an answer. George's daily prayers had given him a sensitive spirit that was in tune with his heavenly Father, whom he often addressed as Mr. Creator. He had made many of the important decisions in his life as he sensed God's guidance during these times.

One morning, George asked God why He made this universe. In his heart, George heard a reply, "You want to know too much for that mind of yours!" Then George asked Him what man was made for. Again, he heard a reply in his heart: "Little man, you are still asking more than you can handle." George narrowed his prayer, so he asked, "Mr. Creator, why did You make the peanut?" He sensed God's reply: "Go to your laboratory and I will show you."

George hurried to his laboratory, left orders for no one to disturb him and sent for three bushels of peanuts. Sitting quietly behind a closed door, he picked up a peanut, inspected it carefully, and waited for instructions on what to do next. Continuing his conversation with God, he prayed, "What do I do now?"

Again, the answer silently filled his heart: "Take the peanut apart."

With his knowledge of chemistry, George quickly began the process of separating the components of peanuts into sugar, starch, fat, oil, amino acids, and other chemicals, lining them up in beakers across his work bench. He continued to pray for guidance, and the answers continued to come as silent messages. When he asked, "What do I do with the parts?" The answer was more specific.

This time George realized he was being guided to reassemble different substances from the peanuts he had separated. He remained in his laboratory for three days, combining the components under different temperatures and pressures to make new products. He ate peanuts when he got hungry and napped in his chair when he became tired. His excitement about the multitude of new products that could be made from this one food pushed him to work as fast as he could for answers the farmers needed.

When he finally opened the door to the lab, he invited inquisitive students and faculty inside to see the rows of containers that contained dozens of goods made from peanuts and told them the story of how he had been able to discover them.

Peanuts were no longer a worthless crop that only served as food for hogs. George had produced peanut milk and a variety of other foods from peanuts. Other useful items were plastic, ink, bleach, and washing powder, as well as personal items such as shampoo, facial cream, and shaving cream. New products continued to be produced. Each of these products could be manufactured and sold for a profit. George and his assistants made samples of delicious foods that could be made from peanuts to give businessmen. He created bulletins filled with peanut recipes to give farm families. Behind the scenes, George worked with business-men to find ways to produce and promote some of his discoveries. Gradually, businessmen began to market the products made from peanuts, and peanut farm-ing became profitable.

An awareness of George's intimate relationship with Mr. Creator spread across the cam-pus. Students who were hungry

George Washington Carver, 1910

57

to understand the Bible persuaded him to teach a weekly Sunday night Bible study. The study, which began in 1907, became one of the most popular events on campus and lasted for thirty years. A lifetime of living by faith had given George a wealth of godly life lessons and an opportunity to introduce many students to his friend Jesus, who gave him strength to do all things.

Review and Expand Your Knowledge

1. When George received a letter informing him that boll weevils were headed toward Alabama cotton fields, what crops did he recommend that farmers plant and why?

2. For years, many farmers had depended on receiving cash for their cotton crops as a way to make a living. In 1904, due to the boll weevils, there was no profit from their cotton crops. For the farmers who had ventured to add other crops, especially peanuts, there was a little food to eat. Did the peanut crops provide an adequate source of cash for the farmers that year?

3. What three things did George do to help the farmers sell their new crops for a profit?

4. What was George's inspiration for discovering many valuable chemicals and products from peanuts?

5. As the story about George's inspiration for finding new products from peanuts spread across the campus, what extracurricular study did a group of students ask George to lead? How long did this study continue?

CHAPTER 13

PARTNERING WITH GOVERNMENT AND BUSINESSES

George was now becoming well known and respected across America. He received many opportunities to work for corporations and to be the featured speaker at large conventions, but his desire to help farmers was always a priority. His notoriety did not prevent him from finding time to talk with small groups of people who worked outside on farms where he shared ways to improve crop production, as well as his deep beliefs about creation and Mr. Creator.

Boll weevils had convinced many farmers not to rely only on cotton crops to make a living. Many farmers had begun to experiment with other crops. George continually tried to think of ways to get poor farmers to listen to the solutions he could provide to enable them to have more prosperous farms. One of the things he tried to convey to farmers was the value of finding ways to recycle and reuse things they owned. When George received a telegram informing him of the death of ninety-eight-year-old Moses Carver, he imagined how proud his Uncle Moses would be that the same lessons George had learned about recycling were being passed on to a new generation of young people and even to seasoned farmers.

Galena Man, 98, Was Friend of Lincoln

Moses Carver, Pioneer, Dies at Home of His Son — Knew Martyred President in Springfield, Ill.

The Globe Bureau,
308 Main Street.

Galena, Kan., Dec. 19.—Moses Carver, 98 years of age, neighbor and friend of Abraham Lincoln, died yesterday morning at 3 o'clock at the home of John Carver, south of Galena. The body will be taken to Diamond, Mo., this afternoon, where interment will be made beside his wife, who died twenty years ago.

In his boyhood Carver made the acquaintance of Abraham Lincoln through his residence in Springfield, Ill. Born in Dayton, O., Carver and his family moved to the home of Lincoln in his boyhood, where he remained until he was 27 years old, in 1839. Ten years before the exodus of the "Forty-niner" he, with his wife, came to Diamond, Mo., and homesteaded, where he remained until four years ago. Becoming too old to look after his home, Carver came and joined his family south of Galena.

Although nearly 100 years old, he was quite agile, and looked twenty years younger than he really was. He took considerable pride in relating the scenes of his early life, especially those in which the martyred president figured.

When World War I broke out in Europe, President Woodrow Wilson tried to stay out of the war. However, Germany's aggressive actions left him no choice. After the Germans torpedoed and sank a passenger ship, the *Lusitania,* causing the deaths of one hundred and twenty-four Americans, the United States joined the war effort with the Allies in 1917. Immediately the U.S. was cut off from a critical supply of aniline dyes, which were essential for printing newspapers, dyeing army uniforms, and making paints for ships and cars. All aniline dyes were produced in Germany and their production became a closely guarded secret when the war started.

George's commitment to serving his country caused him to shift much of his time to doing scientific research. He made it one of his missions to use his research skills and find new sources of the important dyes. For over a month, he collected as many promising plants as he could find. He then locked himself in his laboratory and began a search for dye colors. Amazingly, he discovered 536 colors of dye and then wrote to various industrial leaders, inviting them to use and manufacture his discoveries, which were made from readily available raw materials. To demonstrate to his students that dyes could be produced from readily available resources, George always wore a brightly colored tie that he had made and dyed himself.

In another of his war projects, George developed methods to dehydrate foods for use on ships bound for Europe as an alternative to sending fresh fruits and vegetables that tended to rot on the trip. He also promoted the use of sweet potato flour to make bread.

Over the next several years, George continued to conduct research on peanuts, sweet potatoes, chickpeas, soybeans, and other crops that could be sold for a profit, as well as provide food for families. He discovered hundreds of products, many of which later became items that were sold in stores. George's speeches and research demonstrations were widely-reported as George was interviewed by several publications. Even the famous Thomas Edison contacted George about taking a job in his laboratory. Although briefly tempted, George knew his work would always be at Tuskegee.

George soon became a popular speaker and had a busy travel schedule, often speaking before influential committees and government officials. On one occasion, he spoke to the United Peanut Growers Association in Atlanta, Georgia. When George opened his suitcases and laid out many of the products made from peanuts on the table, he had the full attention of his audience. His speech was so inspiring the committee voted to make him their spokesman to request that Congress put tariffs on peanuts that were being imported from Asian countries and sold at cut-rate prices. A tariff could keep the price of homegrown peanuts at a level where farmers could make a fair profit.

On the morning of the Congressional hearing in Washington, D.C., to ask for tariffs on peanuts from Asia, George walked into the room lugging a huge suitcase. The chairman was obviously unimpressed with the old Negro man wearing a suit that had been mended several times with a weed in his lapel and wearing a brightly colored red tie. But sticking to the schedule, he called Professor George Washington Carver to the front and gave him ten minutes to speak.

George began pulling bottles and boxes from the suitcase and lining them up across a table, talking as he demonstrated products made from peanuts. When several members of the committee seemed to be bored by what he was saying, he began to talk about a perfect meal that could be made from peanuts and sweet potatoes. One of the congressmen interrupted to ask, "Do you want a watermelon to go along with that?" The room immediately became silent following the obviously insulting question, as everyone knew white people sometimes referred to watermelons as a favorite food enjoyed by illiterate blacks. George felt a flush of anger, but he did not take the bait. He replied that a dessert would be nice but not necessary, as the recent war had taught them that. The committee laughed, and George resumed his talk and demonstrations.

From that point on, he had the full attention of the committee. He proceeded to display the many amazing products he had made from peanuts as he explained their potential. When his ten minutes was up, the chairman extended his time. He finally concluded an hour and forty-five minutes later as the entire audience honored him with loud applause and cheering. A week later, Congress put large tariffs on Asian peanuts that allowed American peanut farmers to receive fair, competitive payments for their crops and not be outbid by Asian markets.

All of these projects and speaking engagements gave Tuskegee much credibility and helped to improve the school, but George wanted to make a difference for the farmers, both black and white. Many farmers were making profitable changes, but many others continued to live under terrible conditions that could easily be improved. George had answers, but they involved reducing or abandoning cotton crops. How could he get them to listen and bravely make changes when local farmers had only planted and harvested cotton for decades?

Review and Expand Your Knowledge

1. What problem finally convinced many farmers not to rely only on cotton crops to make a living?
2. What crops other than cotton did farmers begin to grow and rotate from year to year?
3. What did George wish his surrogate father, Moses Carver, could have seen before he died?
4. When did America join the Allies in World War I? What were two ways in which George helped contribute to the war effort?
5. In the midst of a talk before a Senate committee, one of the senators made an insulting remark to George. How did George respond? In your opinion, what would have been the results if George had responded angrily?

6. What did George do to help southern farmers get a fair price for their peanuts in America and not be undercut by cheap peanuts sold by Japanese and Asian farmers? What is a tariff?

7. What advice did George give to help farmers have more productive crops? Why did some farmers choose not to take his advice?

8. What was one way George researched peanuts to discover new products?

9. Why did George advise farmers to reduce or abandon growing cotton?

CHAPTER 14

A PANORAMIC VIEW OF THE CONTRIBUTIONS OF AN AMERICAN HERO

Recently, members of the U.S. Department of Agriculture climbed aboard a small plane for an aerial tour of the farms surrounding Tuskegee. It had been almost 120 years since George Washington Carver had shown members from the Department this same area from a mule-pulled wagon where thousands of acres of cotton had been devastated by an invasion of boll weevils around 1904. This time, the view from the plane revealed scenes of thriving farms—acres of green corn, white fields of cotton, fields of peanuts, and a variety of other cash crops, all ready for a profitable harvest. There had been many amazing changes since George began working at Tuskegee, where he used his vast knowledge of plants to help southern farmers overcome years of deep poverty and learn ways to make farming a prosperous occupation.

During the late 1930s, after forty years of working at Tuskegee, George was able to see his vision for improvements in farming unfolding. The farmers who had taken his advice to add new crops such as peanuts, soybeans, chickpeas, and sweet potatoes to the traditional cotton-crops-only finally emerged from the crushing poverty of the Great Depression. George had introduced crop rotation, natural fertilizers, and legumes, which put nitrogen back into nutrient-depleted soil, and other farming methods that had enabled farmers to produce healthy crops and make a decent living. In spite of the uncertainties of World War II, farmers finally had reasons to hope for a prosperous future. The tour guide recounted that when George was nearing the end of his life, his vision of the future was a time when farmers were not merely surviving but were thriving, respected citizens living in a country where civil rights were guaranteed to all by a Constitutional Republic. Throughout the more than forty years of working at Tuskegee, George had sought to follow God's guidance and fulfill God's purposes for his life as he turned down a number of lucrative job offers and chose to remain at Tuskegee.

Many farmers had finally accepted answers and solutions George had given them, even though many had resisted his good advice at first. His efforts to get southern farmers not to plant cotton year after year on the same soil was an ongoing battle. Cotton was the only crop most knew how to plant, harvest, and get to a market that paid cash. The arrival of the dreaded boll weevils in the early 1900s forced farmers to make major changes in their traditional farming practices. At first, only a few courageous farmers were willing to take George's advice and plant peanuts and other legumes instead of cotton. The new crops were a source of food and had the added benefit of providing a solution to the depleted soils where cotton had been planted for years. Legume plants were sources of nitrogen, which had been depleted by continuous cotton crops. By plowing under the roots of the plants, cotton could continue to be successfully grown in good soil as long as the crops were rotated periodically.

Later when peanuts were overproduced and piled up in warehouses, George worked hours in his lab and discovered over 300 useful products that were made from peanuts and could be sold commercially. When the price of peanuts was being undercut by cheaper prices from Asian farmers, George met with the Congressional Ways and Means Committee and demonstrated many of the useful products made from peanuts and persuaded them to raise tariffs on Asian peanuts to allow American farmers to sell their crops at a profit.

By the early 1940s, peanut farming had become one of the most profitable crops in the South. George discovered hundreds of uses of peanut products. Five million acres of farmland now produces over $500,000,000 in peanut production. At one time peanuts were almost worthless as a cash crop, used primarily as a source of animal feed. Sweet potatoes, pecans, soybeans, cow-peas, wild plums, and okra were also popularized by George as they became other valuable cash crops for farmers. Although large crops of corn, wheat, rice, and rye were grown more in northern states, they also did well in the South. Former black slaves and poor white farmers had become prosperous farmers by following George's advice.

❁ ❂ ✛ ❄

During an earlier visit to the Agriculture Committee in Washington, George received national recognition for making flour from sweet potatoes. Some of the other useful products he produced in his lab were for the hundreds of products he had made from once almost worthless crops, such as soybeans and chickpeas. Other contributions were improved kinds of food plants, dyes of different colors, fruits and vegetables that could be dried and could make long trips across the ocean without becoming rotten, and improved kinds of food plants.

Trying to find ways to improve race relations between white citizens and black citizens was always important to Booker T. Washington,

President of Tuskegee. He thought the best way to destroy racial prejudice was for black people to make themselves a useful member of the community in which they lived, rather than encourage forceful or violent protests. George's life was a perfect demonstration of this idea.

Then in 1915, George's friend and boss died after a short illness at the age of fifty-nine. This was devastating to George. Many white people attended his funeral, including former President Theodore Roosevelt, who insisted on reading the eulogy. After the funeral, Mr. Roosevelt sought out George. They walked around the campus and discussed the experimental farm as Mr. Roosevelt encouraged George in a profound way, ending their conversation with the comment, "There is no more important work than what you are doing."

When the Great Depression began in 1929, George found even greater opportunities to help others. Thousands of job opportunities disappeared, and many people worried if they would have enough food to eat. By 1932, over fifteen million people had lost their jobs. Wages were also cut and hundreds of banks failed.

Many small farmers became self-sufficient and produced enough food for their family to live on, but many other people barely survived that year.

George wanted to do more to teach farmers how to survive the difficult economic times. He especially wanted to show them that farming could be a prosperous career. In 1935, Fredrick Patterson became the new president of Tuskegee. By then, George was seventy-one and in declining health. Patterson hired Austin Curtis Jr. to be his assistant so he could continue limited amounts of traveling and speaking.

On the fortieth anniversary of George's arrival at Tuskegee, the Institute held a large celebration in his honor. Many reporters came from *Time* and *Life* magazines, and a variety of newspapers and radio stations also sent people to cover the event. Reporters were fascinated by his life. They wanted to know how he had managed to get an education and why he had decided to come to Tuskegee. The new president of Tuskegee thought this was a good time to establish a museum on the campus about George's life story. Agreeing to support the idea, George thought the challenges he had faced would be a way to encourage other people struggling to have a better life and live useful lives. It might also provide an opportunity to show farmers proven ways to become self-sufficient and to produce crops they could sell. Realizing he did not have many years left, he continued to plan the museum, along with a room that contained his artwork.

During this time, George only accepted a few speaking engagements, because they were very tiring for him. There was one conference, the Chemurgy Conference in Dearborn, Michigan, where he made a special effort to attend as the speaker. The famous inventor, Henry Ford, brought top agricultural scientists and industrial researchers together for the conference. As the most popular speaker there, George left attendees with practical ideas for using products farmers could provide. Following the conference, Henry Ford himself met with George and invited George and his assistant to his home in Ways, Georgia, where they continued to develop some of George's ideas.

The day came when the museum opened. George personally greeted each of the thousands of visitors, black and white, as people recognized the enormous impact of his life. He once told a reporter, "I am not a finisher; I am a blazer of various trails. Little of my work is in books. Others must take up the various trails of truth and carry them on." With that philosophy, he made the decision to establish a George Washington Carver Foundation and donated more than $32,000 of his own money to help young black men and women carry on the work he had begun. This money had been gradually saved throughout years of frugal living. Part of the money was used to attach a modern research lab to the museum so students could work on their own research ideas.

During the next few years, George received many awards and honors. Schools were named after him. He received honorary memberships in prestigious science organizations and was awarded honorary doctorates from Simpson, Selma, and the University of Rochester. He received letters of appreciation from all over the world. His friends and admirers included Mahatma Gandhi, U.S. presidents, senators, representatives, and cabinet members.

Nothing pleased George more than seeing young black men and women following in his footsteps and knowing his life had been given purpose by the Creator. In a radio interview not long before his death, he said, "Why I should be so singly honored is more than I can figure out. I have just endeavored to do my little bit in the world as fast and as thoroughly as the Great Creator of all things gave me the light and strength."

When Booker T. Washington invited George to give up money, position, and fame to come to Tuskegee, he challenged him to engage in the hard work of helping people rise from degradation, poverty, and waste to their full human potential. Knowing this was God's call to him, George walked the paths where God led without regret and lived to see a bountiful harvest in the lives of multitudes of students and farmers he had influenced.

Review and Expand Your Knowledge

1. What were some of the events that increased George's popularity and fame during his first thirty-nine years at Tuskegee?
2. Americans learned more about George during a celebration of his fortieth year at Tuskegee. What kind of publications wrote about his life?
3. What was the purpose of the museum that was established a few years later? What did George consider his greatest accomplishment in life?
4. In your opinion, do you think George regretted not accepting one of the prestigious job offers that paid a top salary, included an up-to-date laboratory, and held a position of authority? Why or why not?

5. Where did George obtain the $32,000 he gave to a George Washington Carver foundation?

6. The year 1904 was a disaster for many southern farmers. Notice the change from "120 years" to "century." What happened in that year and how did that year compare to farms a century later?

7. What did George do to help farmers make a living with their peanut crops (two ways)?

8. What happened in 1929 that caused Americans everywhere to worry about having enough food to eat?

9. Name at least four famous people from around the world who were friends with George or who wanted his advice.

THE SURROUNDING STORIES

Beginning and Growth of the United States of America

As soon as European nations heard that Christopher Columbus had discovered a new route across the Atlantic Ocean to India and the Far East in 1492, several nations began taking the dangerous trip across the Atlantic to investigate his intriguing discovery. They intended to establish trade agreements with countries like India and China. However, they soon discovered this land was not the Far East, which the explorer Marco Polo had reached earlier by land. It was a huge land mass, surrounded by two oceans between Europe and the Far East. These lands, previously unknown to Europe, were referred to as the New World.

The right to claim ownership of this land resulted in numerous battles. For centuries, Spain, England, France, Portugal, native Indian empires, and other countries fought each other for the right to claim land in the New World. As news of the opportunities spread to Europe and other countries, settlers began to arrive looking for the land of opportunity. Some came in the name of the king of their former country; some just came looking for a better place to live where they could speak freely about what they believed, and farm their own land. Most importantly, they were no longer under the rule of tyrannical kings.

Some of the early settlers made their homes along the eastern Atlantic seacoast, and were in an area claimed by England. In 1775, several battles were fought with England over the ownership of the land and England's right to impose taxes on the settlers. Then in 1776, the original thirteen American colonies sent the famous Declaration of Independence to England, declaring themselves the independent sovereign nation of America. A ragtag army of American patriots fought a war with the mighty army of England before the independence of the first thirteen American states was achieved and a peace treaty with England was signed.

❀ ✪ ✛ ❄

A New Government after the Revolution

The next obstacle facing the new nation was getting all thirteen states to agree on a plan for how to govern itself. The first plan to govern the thirteen states was known as the Articles of Confederation. Fearful of a strong central government, all the states tried to maintain their own state's rights, which were no more than existing as thirteen separate nations. The proposed Articles of Confederation were finally agreed upon in 1781, but the weak concessions the states had made left little power for a centralized government, and numerous problems arose that simply could not be reconciled. It became obvious that the Articles needed to be replaced by a new plan with a strong centralized government which all thirteen states could agree upon.

Delegates met in Philadelphia on May 25, 1787, to hammer out a constitution in which the rights of all the citizens, rich and poor alike, were guaranteed. It required each state to give up some of their rights in order to have a stronger government. All summer the delegates discussed and rediscussed the issues. They wrote multiple drafts, as they came to realize they were not merely amending the old Articles. They were drafting an entirely new kind of government that none of them had ever seen in operation before. They rejected all forms of kingship and special rights for the rich. Rather, they envisioned a government by which all citizens would choose their local, state, and national leaders by voting for them. The Constitution listed the rights that all citizens were guaranteed and would live by. It was called an experiment, because the founding fathers had never actually lived under a Constitutional Republic. The Constitution was officially ratified in 1788.

The United States Constitution described the three branches of government and how they would be set up, as well as other specifics about governing the nation. A method for amending the Constitution was included. The heart of the Constitution is the Bill of Rights that enumerates the basic rights each citizen is guaranteed. First Amendment rights guarantee freedom of religion, speech, the press, assembly, and petition. Other rights are also spelled out in the Constitution.

❀ ✪ ✛ ❄

Legal Rights of Slaves

One of the most divisive issues from the beginning of the United States was deciding if all the Constitutional rights applied to the slaves. Slavery was already established as a big part of the economy of the southern states when the Constitution was written and adopted. Slavery was often debated in Congress, but meaningful laws had not passed, except for the Slave Act of 1807. The purpose of this act was to prevent slaves from being imported into the United States from Africa.

Many of the founding fathers and congressional senators and representatives from northern states wanted to end slavery, but suddenly freeing 700,000 slaves without a source of food, homes, or income would create enormous problems. Members of Congress also understood that the pro-slave states would withdraw from the nation if slavery were to be abolished. Maintaining a united nation was a major reason for hesitating to make slavery illegal. Although a slave owner himself, President Washington thought carefully about the future of slavery and looked for ways to gradually end slavery in America. His will decreed that all of his own slaves were to be freed at his wife's death. He also left ample provisions to care for his elderly and sick former slaves, and he provided for the education of the children of his slaves.

Washington hoped that similar methods could be extended to all slaves in America to eventually end slavery. However, the number of slaves increased significantly soon after the war to almost four million. One reason for the growing slave population was the invention of the cotton gin, enabling cotton farmers to produce bigger cotton crops that needed more slave workers. As more time went by, the Deep South became a cotton-based economy, dependent on slave workers. Cotton was creating a great deal of wealth for big farm owners.

Nevertheless, the bold words of the Declaration of Independence, "We hold these truths to be self-evident, that all men are created equal, that they are endowed by their Creator with certain unalienable rights that among these are Life, Liberty and the pursuit of Happiness…" were genuine beliefs of most of the framers of the Constitution.

❁ ✪ ✛ ❋

Slavery before the Revolutionary War

The practice of slavery can be traced back thousands of years. Since ancient times, powerful armies clashed in battles to gain territory or subdue other nations thought to be a danger or a competition to them. The victors often forced the losers to serve them as slaves. Throughout history, slaves came from a variety of places, but during the past few hundred years, most were from Africa. From the early 1500s to about 1850, there were several major slave trade routes from Africa to destinations both to the east and to the west. Some historians estimate that Arab slave traders took millions of African natives to Asia as slaves. Slave traders, mostly from Portugal, shipped another 11,000,000 African natives to the Western Hemisphere, of which only about 5% were taken into areas that are now part of the United States. By the time the American Constitution was adopted, slavery already existed in all of the original thirteen states. After the Revolutionary War, the northern states rapidly became an industrialized section of the country with little demand for slaves, while the southern states remained a farming section where slaves were only used on plantations and large farms. Most white farmers in the South did not own slaves, because their farms were small enough so that family members could grow their own crops and get all the work done themselves.

The Carvers' situation was very different from that of other slave owners. They purchased George's mother as a slave, even though they did not agree with the idea of slavery. They had been unable to find local helpers to hire, because the young people had left due to the war. Mary and the Carvers always had a good relationship that developed into a real friendship. The Carvers raised Mary's children as their own children after she was abducted by outlaws.

Before the Civil War began, all of the northern and western states were committed to either gradually ending or outright abolishing slavery in their states. These decisions were not controversial because there were only a few slaves in the northern states at this time, with many jobs available in the factories or other industry-related positions. Public opinion in northern states gradually, but increasingly, favored abolishing slavery. By 1861 when the Civil War began, slavery had not been abolished in any of the southern states.

❀ ✹ ✛ ❄

What Was behind the Disastrous Beliefs in Superior and Inferior Races?

Slave traders consistently told their "customers" that black Africans were inferior to white races, promoting a long-lasting stereotype that they were born with a low intelligence and probably were not fully human. This was never true, but the lies persisted for generations. Perhaps portraying black African natives in this way made slavery more acceptable to those who had moral objections to buying, selling, and owning human beings. If this were true, then owning "soulless" farm workers was no worse than owning farm work animals. When a lie is heard over and over, it tends to eventually be accepted as true. The stereotype that members of the black race were intellectually inferior to the white race was widely believed by many people throughout America at one time, especially in the South.

There were several reasons why the lie was made more believable. One reason was that the culture of people from central and southern Africa was distinctively different from the culture of other people who came to America from Europe. African slaves had not developed systems by which they taught their children to read, write, or do mathematics. Nor had they developed tools made of iron and steel, such as steam engines, rifles, or farm tools. Superstitious beliefs in evil spirits dominated both the medical and spiritual practices of the slaves.

The Sahara Desert in northern Africa covers an area the size of the entire United States. It was a natural barrier that had prevented most native Africans from migrating north of the desert, where they could have learned about new ideas and technologies that had developed in other cultures along the Mediterranean Sea. African natives were able to build canoes, but large seaworthy ships had not been developed by them to explore long distances. The desert barrier was a primary reason why other countries did not often venture into interior African territories. This left a large swatch of the African continent that was isolated from learning about new technologies or being influenced by European cultures and technologies.

❀ ✿ ✛ ❄

The Industrial Revolution and an Agricultural South

Years before the United States was carved out of the New World as a sovereign nation, a large number of slaves were already living in North and South America, the Caribbean Islands, and in many other places around the world. Although there were strong feelings and beliefs about the morality of slavery from the time when the first slaves were brought to Virginia in 1619, the practice continued to expand as the demand for cheap manual labor grew. There were no national laws making slavery legal and none making slavery illegal.

Northern states took advantage of the many streams and rivers available to help build their economy. Textile factories used water-powered or steam-driven engines as a source of energy for their products. Water wheels were used both in producing spools of thread and in turning the looms that quickly wove the cloth. Northern and southern states worked together for years as each side profited economically from a steady supply of cotton from the South that provided the raw materials needed to make fabrics.

Northern states also capitalized on the need for additional industries. Trains and ships built in the North transported the fabrics throughout many places in the U.S., and even to Europe. Ship building and shipping became another prosperous industry in northern states.

The demands for nails, screws, rivets, building materials, tools, engines, and machines were pressure points that increased the need for iron and steel. The northern states produced large quantities of iron and steel products, in addition to textiles and ships. Factories, goods, and methods of transporting these items built a manufacturing center, while the South remained primarily an agricultural center.

❀ ❂ ✛ ❄

The View of Slavery from the North and the South

At first, in the new nation of America, there was no widespread feeling in the North that slavery was wrong. The North had taken advantage of new inventions and discoveries to build a variety of factories and manufactured goods, with a minimum need for slave labor.

Attitudes about slavery began to change when the book, *Uncle Tom's Cabin,* was published in 1852. It was widely influential in how people in the northern states viewed slavery. After reading the book, many people began to believe all slaves were mistreated and lived miserable lives. They believed that all slave owners were like the fictional character Simon Legree in the book who treated his slaves with extreme cruelty. Whether or not to free the slaves became a passionate, divisive issue in both the North and the South, but each side had different reasons for their beliefs about freeing the slaves.

For the most part, slaves in the South were treated decently and provided with clothes, food, medical attention, housing, and at times, social events. Relationships between slaves and owners were often friendly, even though slaves could expect harsh punishment for rebellious actions and disrespect. However, slaves were often considered more like a financial investment than as fellow human beings made in the image of God. Abolitionists, mostly from the North, noted this well and filled the newspapers with passionate anti-slavery editorials.

Americans from all sectors of the country had to decide the answers to some hard and important questions about African slaves. Were slaves merely a piece of property or were they fellow human beings? Were they made in the image of God? Were rights given to them by God or were they given by the government?

❀ ✪ ✤ ❄

Fears of Freeing All Slaves

Many white farmers from southern states who had never owned slaves objected to freeing the slaves, because of fears of how they would survive as the number of slaves in America increased to almost four million. The memory of Nat Turner's attempted rebellion caused many white families to fear that former slaves might resort to violence to obtain the things they needed to survive.

Turner, a slave himself, had felt it was his calling to start a rebellion of slaves several years earlier in 1831. With a few followers, he murdered his owner, his owner's wife, and all their children. He then went on to murder about 60 other white slave owners. In the aftermath of the uprising, the rebel slaves were caught and executed, but the incident left a lingering fear of what might happen if all the slaves were suddenly freed.

Another attempt to start a rebellion of slaves in 1859 by John Brown and his followers interjected more fears and emotions into the debate. John Brown, a radical white man from the North who hated slavery, assumed that many slaves would join him and his army in an armed revolution if they could acquire enough weapons. To get the ammunition, a government arsenal at Harpers Ferry was captured. In the end, the rebels were captured, and Brown was hanged. The incident was widely reported in northern newspapers, the death of John Brown became the words in a popular song, and the slavery debate heated up even more.

❀ ○ ✛ ❄

The Civil War and Reconstruction

The Union had held together for almost a hundred years after the Revolutionary War as compromises on slavery were made to admit one new slave-holding state simultaneously with one slave-free state. As more states west of the Mississippi River sought to be admitted as slave-free states, it became obvious that this would soon undo the balance of pro-slave vs. slave-free states, so that Congress would inevitably make slavery illegal throughout the nation. It was also obvious that many state laws were going to be overridden by federal laws. In 1861, the southern states declared themselves to be an independent country and withdrew from the United States, forming their own government and making their own laws and printing their own money.

Fighting between the Union army and the Confederate army soon broke out. Both sides were confident they would be able to win, and the war would be over soon. However, it lasted for four long, bloody years before a devastated South finally surrendered. George was born during the Civil War in approximately 1864. Abraham Lincoln had already signed the Emancipation Proclamation, so as soon as the war ended, all slaves were freed.

During Reconstruction, the South was divided into military districts under the rule of Northern army generals, who enforced new rules on the South passed by a group of Republican representatives in Congress, known as the Radicals. President Lincoln had hoped to promote a spirit of kindness and charity between the two sides, but the Radical Republicans were more interested in punishing white people in the South than in achieving a lasting peace between the two sides. Nevertheless, as soon as the Northern army left the South, the Reconstruction became even worse for former slaves. With the assassination of President Lincoln in 1865, his wise leadership was absent during the difficult days of the Reconstruction

The economy of the South was in ruins after the war. With most of the fighting taking place on southern soil, farm land had been ravaged; livestock slaughtered or stolen; and crops taken or destroyed. The Northern army burned many plantation homes and other structures to the ground. Without slave labor, even the families that escaped the burning of their plantations could not continue to plant or harvest their crops and were forced to sell as much as they could.

The first years of Reconstruction were a disaster to the southern states as efforts to rebuild homes and farms were confounded by incompetent or even corrupt government authorities. Large numbers of northerners moved south during the Reconstruction. Some came with a genuine desire to help by building schools, serving as teachers, and helping rebuild essential structures. Others came and took advantage of the unstable conditions in order to find ways to benefit themselves financially and politically.

Living with devastating economic conditions, the South limped through the Reconstruction years. Black and white farmers alike faced widespread hunger and poverty. By the time George had accepted a job at Tuskegee in 1896, conditions for small southern farmers, both black and white, had not improved over the past thirty years when the Civil War ended. Farm families were struggling to just survive.

Following the war, most white citizens were able to vote, but they voted only for Democrats for many years. A few black citizens were able to vote after a Supreme Court decision declaring poll taxes and reading tests illegal requirements to vote. Virtually all black voters backed Republicans, but within the past few years this pattern had changed.

An even worse development was that a number of white men joined violent groups like the Ku Klux Klan, who became self-appointed guardians of their communities. They were determined to discourage former slaves from voting or running for political offices or acquiring authority in any area by means of fear and intimidation and even violence. Their tactics were also designed to intimidate white people who disagreed with their harsh methods.

George's encounter as a youth with the two bullies came from an assumption that all dark-skinned people did not have enough intelligence to learn to read. Unfairly accused of stealing his books from a white child, the bullies proceeded to hit and kick him to "teach him a lesson about stealing from white people." White people who saw what was happening were appalled, but turned their heads and walked away in disgust. This was an example and a forerunner of the dangerous racial beliefs that were developing, mostly in the South, but also found in other places throughout the nation. This was one example of the obstacles that George had to overcome as a black man.

❀ ❁ ✛ ❄

Origins of Beliefs in Inferior and Superior Races and Further Consequences of the Lies

Although the assumed belief that black people lacked higher levels of intelligence was never true, many Americans from both northern and southern states believed the lie. These assumed beliefs were later given a new level of credibility by Charles Darwin's theory of evolution. First introduced into European academic circles, Darwin's new explanations about the origins of living things and the novel idea that all animals had evolved from other animal ancestors over long periods of time were a source of fascination to professors, doctors, educators, scientists, and others. Darwin's book, *The Origin of Species,* published in 1859, proposed that all species of living things had evolved from other simpler species. His later book, *The Descent of Man*, hypothesized that humans had also gradually evolved from simpler species of animals, leaving some races superior and others inferior.

"Experts" of human evolution believe that all human races evolved from ape-like animals. During the process of evolutionary changes and the struggle of the fittest to survive, some human races evolved as more superior than others. A common belief was that people who came from lands with harsh cold temperatures were more likely to evolve into highly intelligent races than people who evolved in hot climates.

During the late 1800s and early 1900s, people with dark skin color were assumed by many people to belong to an inferior race. White supremacy, segregated schools, unfair Jim Crow laws, and eugenics policies grew out of these assumed false beliefs about different races. White supremacy beliefs have had devastating effects on the lives of black people.

Darwin's ideas about evolution were debated in academic circles in Europe throughout the late 1800s. But, by the early 1900s, evolutionary theories began to impact the cultures, schools, and world views. White supremacists added some of Darwin's teachings to their arsenal of ideas that supported the beliefs that whites were members of a superior race and blacks were members of an inferior race. George had brief encounters with converts to evolution and naturalism who accused him of using unscientific methods and holding unscientific beliefs, but his strong faith in God and the Bible remained unshaken.

❀ ◉ ✛ ❄

Hitler's Social Darwinism and Eugenics Policies

Adolf Hitler was the first world leader to make Social Darwinism and eugenics part of the official policies of his government. Darwin seemed to have opposed slavery, but that did not prevent others from using his ideas for their own purposes. The research of several recent scholars and historians (Dr. Richard Weikart) have established that Hitler's racial views, along with his eugenics and Social Darwinian policies, were directly related to Darwinian evolutionary views.

Studying methods used by race horse owners to breed thorough-breds inspired Hitler to try to apply evolutionary methods to humans to produce a superior race of humans, known as a "master race." After becoming the chancellor of Germany, he established a brutal eugenics program to prevent intermarriages between the so-called superior and inferior races and adopted even more brutal methods to keep the "inferior" races from growing any larger. Based only on his own prejudices, Hitler concluded that the Jewish race was the most inferior race on earth. He decided the only lasting solution to keep them from corrupting the earth was to completely get rid of them. Before Hitler committed suicide in 1945, approximately 6,000,000 Jews had died or been murdered by his orders. This period is known as the Holocaust. His goal to "rid the world of Jews" has been condemned by almost every civilized nation since.

The Aryan (white) race was regarded by many Germans as the most superior race in the world, and the Jewish race was regarded as the world's most inferior race. Under Hitler's leadership in the 1930s, Negroes, as well as Slavs, Mongolians, and some other races were also classified as inferior races who needed to be controlled later on by his racial policies. At one time, the basis for many of the beliefs about superior and inferior races were promoted by some of the most prestigious scientists, medical doctors, and educators in Germany. Unfortunately, the misguided ideas of superior and inferior races also found its way to America and propped up the idea of white supremacy as true among some sections of America for years.

❊ ✪ ✛ ❉

Martin Luther King Jr.: Breaking through Racial Strongholds

Although George had earned the respect of many scientists, inventors, and political leaders in America, on several occasions during his life, George was the victim of insults and discrimination because of his race.

At one time, strict segregation seemed to be immovable, but George had seen cracks gradually form in this formidable tradition as black and white farmers worked together to deal with their common problems of boll weevils, soils depleted of nutrients, soil erosion, and the Great Depression. They worked together for solutions George introduced, such as crop rotation and other changes in traditional farming methods, tariffs on foreign peanuts, and marketable products made from crops other than cotton.

George did not live long enough to see momentous changes brought about by the Civil Rights protests beginning in the 1950s.

While white southerners were very much opposed to mistreating slaves, a surprising number of white people had accepted the belief that black people were inferior to white people in intelligence and in other ways. The peaceful protests led by Dr. Martin Luther King Jr. during the 1950s and 1960s, supported by faith-filled black churches, exposed the lie, much like Mrs. Liston had done at Iowa State when George was being disrespected because of his race. Her actions and her words forced both the students and the teachers to recognize their own biases based on his race. Before he graduated with a master's degree, he had become respected, recognized, and accepted for his intelligence and character. His many friends at Iowa State were truly sorry to see him leave them when he accepted the job at Tuskegee.

Discrimination of so-called inferior races was not only found in America. It was also displayed on the world stage during World War II where unusually cruel treatment of Jews exposed the evil of hating people based on their race. It was a turning point for how many Americans began to change their beliefs about racial lies.

However, George's lifetime vision to help the struggling poor farmers in the South was about to come true. Farming was becoming a respected and thriving occupation. Without witnessing civil right victories first-hand, he already knew in his spirit that the lies about white supremacy were soon to be rejected by mainstream Americans. George died in 1943, just a few years before the start of the Civil Rights movement of the 1950s – 1960s.

Dr. King lived to see some big changes that occurred after the unanimous 1954 Brown v. Board Supreme Court decision declared racially segregated public schools unconstitutional and the 1964 Civil Rights Act prohibited discrimination on the basis of race, color, religion, sex, or national origin in hiring and firing practices. However, he did not see the final victories either. Like George Washington Carver, he knew his dreams for equal justice and opportunities would one day be a reality.

❀ ❂ ✛ ❄

The Future of Racism in America

Dr. Alveda King, the niece of Dr. Martin Luther King Jr., is a well-known American Christian writer. She quickly discerned the pockets of white Americans who insisted on believing the lie of inferior and superior races, and recognized that this belief was the source of many racial issues and problems in America.

After deciding to follow Christ and adopting a Christian worldview, Dr. King was determined to become a voice for Him to teach others that "There is no white race, no black race, no red race, no brown race, no yellow race, no mixed races. There is one human race." She concluded that we are all of one blood, descended from our original male and female parents, Adam and Eve, who were designed and created by God in His image. George had recognized this truth in his childhood. Throughout his life, George honored God as the creator and designer of everything.

❀ ❂ ✛ ❄

What's Happening Now?

Instead of celebrating real changes in how most citizens have thoroughly rejected former white supremacy racial beliefs, a new group of leaders began teaching divisive racial ideas, such as Critical Race Theory and the 1619 curricula. The two most objectionable teachings in Critical Race Theory are: white children automatically acquire white privilege at birth, and black children are destined at birth to be victims of discrimination. It seems that as "white supremacy" is getting booted out in America, "white privilege" is moving in to take its place. Even more alarming, these are basic Marxist tenets of victims and victimizers. Socialism and Communist governments are being popularized around racial issues and are making headways in American schools and in the culture.

George Washington Carver, like Martin Luther King Jr., didn't live to see all the changes in racial views that occurred in America. However, their faith gave them a vision of a time when all Americans would simply go about their daily lives confident in their God-given rights and opportunities as guaranteed in the Constitution without regard to race.

Patriotic Americans are hopeful that schools will be one of the places where racism finally dies. But the future has yet to reveal what will be the effects of teaching children ideas such as Critical Race Theory, Black Lives Matter, 1619 curricula, and correct (or incorrect) American history.

Review and Expand Your Knowledge

1. Refer to the timeline of events before George W. Carver was born through events that occurred after he died. Copy the timeline on a longer piece of paper to make room for more events. Add other historical events, as well as personal family history if you have access to information about things such as births, deaths, weddings, honors, military service, etc. to get some perspective on how you and your family members fit into this period of history.

2. Find as many things as you can in the story that refers to the year 1859. (Include two years before or after 1859.) Add these events to your original timeline.

3. It seemed puzzling to 18th and 19th century Americans why there were great cultural differences between the people of Europe who were the primary settlers of America and the people of Africa, who were brought in as slaves. It was a common belief that the cultures were different because Africans were intellectually inferior. **This was never true!** What were actually the reasons for the **cultural** differences in these groups?

4. Did slavery originate in America?

5. What group of Americans owned most of the slaves?

6. How did the invention of the cotton gin affect slavery?

7. How many slaves were living in America by the end of the Civil War?

8. By the year 1904, many southern farmers were facing a disaster with their cotton crops. What happened in that year and how did that year compare to farms 100 years later?

9. Planting nothing but cotton year after year caused many problems for poor southern farmers when George came to Tuskegee. What problems did this practice cause?

10. What kinds of changes did George recommend for farmers to remedy the problems?

11. What did George do to help farmers make a living with their peanut crops? (Two ways.)

12. George was born during the Civil War. What were the conditions in the South after the war ended? Why were the northern states not affected in the same ways?

13. George also lived during World War I. What were some of the ways in which he supported the war for the U.S.?

14. George died in 1943 as World War II was underway, and America was recovering from the Great Depression. What was the Great Depression? What were some of the ways George had been able to help farmers be prepared for this time?

15. In your opinion, why were actions against slavery not immediately proposed by the First U.S. Congress, even when most of the writers of the Constitution did not approve of slavery?

16. Was the idea that some races were more superior than other races a widespread belief during the time of slavery?

17. Compare the speech by William Lloyd Garrison with the one given by Alexander Stephens. Which man believed slaves were fully human, made in the image of God? Which man believed slaves were from an inferior race? Which man was an abolitionist? Which man believed in "white supremacy"?

    In a fiery speech in 1854, William Lloyd Garrison said, "But, if they [the slaves] are men, if they are to run the same career of immortality with ourselves … then when I claim for them all that we claim for ourselves, because we are created in the image of God, I am guilty of no extravagance, but am bound, by every principle of honor, by all the claims of human nature, by obedience to Almighty God, to 'remember them that are in bonds as bound with them,' and to demand their immediate and unconditional emancipation."

In a speech given in 1861 by the vice president of the new Confederacy, Alexander Stevens said, "Our new government is founded upon … the great truth that the Negro is not equal to the white man; that slavery, subordination to the superior race, is his natural and normal condition."

18. At the same time black citizens in America were victims of racists beliefs, Jewish citizens in Germany were victims of anti-Semitic racist beliefs. Give an example of how the civil rights of both black Africans in America and Jews in Germany were victims of racist beliefs. What was the lie behind these actions?

19. When George transferred to Iowa State University, he was the only black student at the college. He was treated differently than the other students by fellow classmates and by teachers. Look up the word "stereotype" and tell if you think George was being judged by stereotypic beliefs because of his dark skin instead of by his individual character and abilities.

20. Who was Martin Luther King Jr.? Did he promote violent protests or non-violent protests?

21. What did the Civil Rights Law of 1964 accomplish?

22. Some people are promoting racial curricula in schools, such as Critical Race Theory and the 1619 Project to show that racism and white supremacy are still major evils of American culture. In your opinion, is the following statement accurate? Explain your answer. "Although slavery and racism were a part of American history, they were not born in America. However, they did die here." (Quote by unknown author.)

23. What does Dr. Alveda King mean by the phrase "one blood" as used in Acts 17:26?

24. What did Hitler believe about different races? Where do you think he got his extreme ideas that led him to order the murder of millions of Jewish people he considered "inferior"?

25. Choose one of the following: Write a possible radio interview with a character in the story. Use one person to be the interviewer and one person to be the character being interviewed.

TIMELINE

1619   First Negro slaves brought to Virginia

1730   First Great Awakening began

1775   Revolutionary War with Britain began

1776   Declaration of Independence

1787   Constitutional Convention meets

1789   George Washington becomes first President of U.S.

1790   Second Great Awakening began

1793   Cotton gin invented

1830   First successful locomotive in America

1830s   Large furnaces built to change iron ore into iron metal

1852   Publication of *Uncle Tom's Cabin*

1857   Prayer Revival

1861   Southern states' secession from the U.S. Civil War began

1862   Homestead Act passed

1863   Lincoln's Emancipation Proclamation issued

1865   Confederacy surrenders to end Civil War

1865	President Lincoln assassinated
1865	Reconstruction period began
1876	Telephone invented
1877	Reconstruction ended and all troops left the South
1893	First Ford automobile
1904	Welsh Worldwide Revival
1914	U.S. in World War I
1929	The Great Depression began when the stock market crashed
1938	Hitler invades Poland
1941	U.S. joined Allies in World War II

ANSWERS TO CHAPTER QUESTIONS

1. The land surrounding the Carver farm was described as law-less. What does that mean? In your opinion, would you like for your family to live in a lawless territory with no official laws or law enforcers?

Ans. *There was little, if any, law enforcement to call on when criminals violated the law. There were few, if any, judges to hear and order punishments for criminal cases.*

2. Why did the bushwhackers choose to live in this region?

Ans. *They wanted to live in an area where they could steal and take advantage of others without interference from law enforcement.*

3. In 1861, what did some of the southern states choose to do about being part of the Union?

Ans. *They declared that they were withdrawing from the Union and forming their own country with their own laws and government.*

4. What did the bushwhackers plan to do with Mary after they kidnapped her?

Ans. *They planned to sell her as a slave to slave-holders in another state for a lot of money.*

5. What was the Emancipation Proclamation? Who signed it?

Ans. *A proclamation signed by President Lincoln in 1863 stating that all slaves in the Confederate states were free, although it took the 13th, 14th, and 15th Amendments to the Constitution to give every citizen full civil rights.*

6. Which of the following did Moses Carver encourage the two boys to learn—music, art, reading, a strong work ethic?

Ans. *He encouraged them to develop a strong work ethic and not waste time with unnecessary or foolish activities.*

CHAPTER 2

1. What did Moses and Susan Carver do with the two children who had belonged to Mary?

Ans. *They raised them like they were their own children.*

2. Why were there only a few black people in the South who knew how to read and write after the Civil War?

Ans. *Teaching a black slave to read and write was discouraged, except for religious instruction. It was even illegal in some states.*

3. Where did George go when he wanted to find a quiet place to pray and to think about answers to his questions?

Ans. *He would go outside before sunrise and find some woods where he could pray when possible.*

4.   What were some of the practical things about living on a farm that Moses and Susan taught George?

Ans.   *Moses taught him how to safely collect honey from beehives, keeping some and selling some. Susan taught him how to spin wool and flax into thread, make cloth, and sew the cloth into the clothes they wore. George learned to tan cowhides and make leather shoes and how to breed good hunting dogs.*

5.   What led George to find a pocketknife that he really wanted? What did he think this might mean for his future?

Ans.   *One night he had a vivid dream about finding a knife next to a half-eaten watermelon next to three stalks of corn. The next day he actually found a knife just as he had seen it in his dream. He thought that just as he had received a knife he had often dreamed of owning, maybe God was telling him to pursue his dream of learning to read and going to school and He would watch over him on his journey.*

6.   At an early age George began to make plans for his future. What did his plan for his future include?

Ans.   *He planned to go to school and learn to read and write. He especially wanted to read the books that answered his questions about the things he found in the woods.*

CHAPTER 3

1.   Why did George have to walk eight miles to attend school when there was a good school in Diamond Grove?

Ans.   *The school in Diamond Grove was an all-white school and did not admit black students.*

2. Arriving in Neosho with no food or money, how was George able to stay there for two years while he attended school?

Ans. He met Mariah and Andy, who offered Gorge a job helping Mariah with her laundry business in exchange for a place to stay and food.

3. What gift did Mariah give George that he used every day?

Ans. A Bible.

4. Where did Mariah and Andy go every Sunday?

Ans. To an African Methodist Church.

5. How did George travel from Missouri to Kansas?

Ans. He traveled with a black family from the church moving to Kansas with a mule train.

6. George moved from town to town several times while still in his teens. What did he do in each town where he lived?

Ans. He found a job and worked until he had saved enough money to buy books and go to school.

CHAPTER 4

1. Why did the two bullies who hurt George assume he had stolen the books he was carrying from a white student?

Ans. They assumed that black children were not intelligent enough to learn to read.

2. Even if witnesses see someone committing a very bad crime, why is it still illegal for a mob of people to execute that person? What does the Constitution say about this?

Ans. The Constitution guarantees everyone the right to a fair trial, even if there are numerous witnesses to a crime.

3. What is the Constitution of the United States? What is the section of the Constitution known as the Bill of Rights? What are some of the main rights spelled out in the Bill of Rights of the U.S. Constitution to citizens?

Ans. The Constitution describes the three branches of government: the legislative branch (makes the laws); the executive branch (carries out the laws); and the judicial branch (interprets the laws). The first ten Amendments guarantee certain basic rights of citizens, such as freedom of religion, freedom of speech, freedom of assembly, and other freedoms.

4. Why is a court system designed to administer justice equally for everyone an important part of the United States?

Ans. The U.S. court system is designed so that everyone accused of a crime can have a fair hearing in a court where evidence is presented, usually before a jury and a judge, to determine guilt or innocence. Equal justice for all is one of the basic principles in the Constitution. It is a right that is greatly valued by most citizens today.

5. If two grown white men were seen punching and kicking a young black person in the middle of any town in America today, which of the following would most likely happen? (a) Police would be called. (b) Someone would step between them and command them to stop. (c) Most everyone would

ignore what was happening. (d) Most everyone would be too afraid to get involved.

Ans. Most likely, either police would be called or someone would try to stop them. It is unlikely that this would be ignored, even if it took courage to challenge what was happening.

6. In your opinion, which of the following finally established the principle of equal justice for all American citizens from 100 years ago? (a) New laws passed during the Civil Rights era making certain kinds of discrimination against minorities illegal. (b) A widespread rejection of the idea that the white race is the most superior race in the world. (c) Most citizens today support the idea that everyone deserves to be treated fairly and with equal justice.

Ans. Your opinion.

7. In your opinion, would it be helpful to teach that white students are born with privileges that black students do not have, as some schools are choosing to do?

Ans. Your opinion.

CHAPTER 5

1. How did George Carver get his middle name?

Ans. When he didn't get mail from people he had written, he discovered that someone else in town was named George Carver. Slaves were not given official names, so he just chose George W. Carver as his name and when a friend thought the "W" stood for "Washington," he began signing his name George Washington Carver.

2. How did George respond to learning that his brother James had died from smallpox?

Ans. He went out into the woods where he was able to find peace and meditate and seek answers from God.

3. Why was George not allowed to attend Highland College?

Ans. Highland College did not accept black students.

4. Where did George find encouragement after he was denied attendance at Highland College?

Ans. He sat down outside a church where he heard singing and found encouragement and advice from the kind ladies who came out.

5. What was the Kansas Homestead Act of 1862?

Ans. It was a law passed by Congress in which people could apply for 160 acres of public land in Kansas. If they cultivated the land for five years, they could file a claim for the land and it would belong to them.

6. Why did George choose to give up his land claim and instead try one more time to go to college?

Ans. George began the process of cultivating the land in 1886, but his dream of finishing college was a persistent dream and overrode his plans to own the land. He gave up his land claim and decided to try once more to attend college.

CHAPTER 6

1. Whenever George arrived at a new town, he followed his plan to look for a job and a place to stay. He would then save

his money to buy books and pay for other college expenses. What kind of job did he find at Winterset, Iowa?

Ans. He worked as a cook in a hotel.

2. Where did he go on his first Sunday in town?

Ans. He went to the Winterset Methodist Church.

3. Where did new friends at church recommend that he apply to attend college?

Ans. The Millhollands, a white couple from the church, persuaded George to apply to Simpson College, a small Methodist College that accepted students of all races.

4. How did the interview go between George and Mr. Holmes? Why were George's meager living conditions satisfactory to him?

Ans. George was welcomed to the college by Mr. Holmes, but living quarters were a problem. Most young men boarded in local homes and Mr. Holmes thought George would have a problem finding a place that would board him. When Mr. Holmes offered to let him live in a run-down shack at the back of the college, George happily accepted the offer and set about fixing it up. He was so well-liked by students and faculty, he often received anonymous gifts for his cabin—always anonymously, because he would not have accepted them if he knew who provided them.

5. What business did he start to pay for his college expenses?

Ans. He set up a laundry business and did laundry for students.

6. What kind of courses did George take at this college?

Ans. Art, piano, voice, as well as other standard courses.

7. What changes did one of his professors advise him to make that involved moving again?

Ans. To move to Iowa State College and study plants and horticulture, which he already loved, and let art and music be his hobbies.

CHAPTER 7

1. George said goodbye to many friends at Winterset College who appreciated his character and intelligence. How was George accepted at Iowa State College in the beginning?

Ans. None of the students wanted to eat with him or talk with him. He was required to eat his meals downstairs with the janitors and people who cleaned and prepared meals.

2. When Mrs. Liston learned that George was being disrespected at the new college, she paid a visit to him. How did her visit help him find acceptance among the other students and faculty?

Ans. She demonstrated how much she respected George. She told a faculty member that she and her husband considered George like a son. She said he was one of the most brilliant students in Simpson College. Both students and faculty were ashamed of how they had judged George as not being capable of doing college work because of his skin color.

3. Did George become bitter and angry with the ways in which he was first treated by other students and faculty at Iowa State?

Ans. *Although he was unhappy by how he was being treated, he did not allow this to produce bitterness or anger in his heart.*

4. Did George believe all white people had hateful, racial attitudes toward black people?

Ans. *Absolutely not.*

5. What kinds of jobs did George take to pay for his college expenses?

Ans. *Sometimes he set up a laundry business and washed and cleaned clothes for students. He also worked as a janitor, tutored, was a masseur for the football team, and found other part-time jobs.*

6. What new kinds of farming methods and scientific techniques did George learn that would enable him to help farmers in the future?

Ans. *He learned new farming methods and scientific techniques that would increase production and the quality of plants grown by farmers. He learned ways to keep the soil healthy and fertile. He was trained in how to analyze soil to discover important substances such as nitrogen, potassium, phosphorus, and other nutrients that had been depleted from the soil, and he learned practical ways to replenish these substances in the soil. He became an expert in crossbreeding plants.*

7. What kind of jobs did George have while he was finishing his Master's degree?

Ans. He was hired by Iowa State College to stay and earn another degree while teaching students as a professor. He also co-authored two books with one of his professors about treating and preventing fungal diseases in cherry trees.

8. What sad news did he receive before he finished college?

Ans. His beloved guardian, Aunt Susan, died before she had a chance to see him finally achieve his long struggle to finish college.

9. What unique achievement did George accomplish at the university in 1894?

Ans. He became the first African American to earn a Bachelor of Science degree from Iowa State College of Agriculture and Mechanical Arts.

CHAPTER 8

1. When George finished his master's degree in 1896, he was invited to retain his job as an instructor at Iowa State College. He also received several other offers for employment, all of which he declined. What kind of work did George envision for the rest of his life?

Ans. He wanted to use his knowledge and expertise to help poor struggling farmers in the South become more successful as farmers.

2. Who sent him the offer for a job that he was most interested in accepting?

Ans. *He declined some jobs that paid much more money than Tuskegee, but he was most interested in a letter from Booker T. Washington inviting him to serve as the head of the new agriculture department at Tuskegee Institute; there were many poor black students who could benefit from his help.*

3. Think about why George chose to give up a good-paying job at Iowa State College where he was highly respected and could do the kind of research he loved in order to move to Tuskegee Institute, one of the poorest schools in America. In your opinion, why do you think George chose Tuskegee?

Ans. *He believed that working at Tuskegee was where he could help the most black and poor farmers. Other reasons in your opinion?*

4. What salary and benefits did Booker T. Washington offer George to move to Tuskegee as the Dean of the Agriculture Department? What three things did he ask George to give up to work with him?

Ans. *Booker T. Washington offered him a salary of $1,000 a year, his own room in the student dorm, and food at the cafeteria. He asked George to give up money, position, and fame.*

5. How did the Jim Crow laws in the Deep South limit the civil rights of black citizens during this time?

Ans. *During Reconstruction, black people were given the right to vote, and they had elected several of their own people into office. Jim Crow laws were aimed at preventing blacks from voting or running for office. They further eroded many of the civil rights they had been given during Reconstruction. Black and white citizens came to live in segregated conditions under the policy of "separate but equal," even though black people often did not receive equal benefits.*

6. For a few years, George Washington Carver and Martin Luther King Jr. were contemporaries. Without a doubt Dr. King heard of the work of George Washington Carver, but he never had a chance to meet him. Both men deeply desired a time when there would be full civil rights for all citizens. In your opinion, do you think George would have approved of Dr. King's method of nonviolent protests?

Ans. Yes. He had always promoted cooperative, non-violent methods to bring about changes.

7. The only source of income for many southern farmers was cotton crops. Why did George advise farmers to practice crop rotation or change the kind of crops they planted every year?

Ans. Planting cotton year after year in the same soil had depleted important nutrients from the soil, so that farmers were no longer producing healthy plants. Farmers could plant legumes, such as peanuts and soybeans, in the soil and restore the nutrients cotton plants needed. He also recommended the practice of crop rotation to prevent further depletion of nutrients that resulted from planting cotton over and over in the same fields.

8. What conditions did George find on the school campus when he arrived at Tuskegee?

Ans. Classes were held in run-down shacks. The campus was in a swamp. There was almost no equipment for the labs and classrooms. *There was little money allocated for Tuskegee from the State.*

9. What conditions did he find on most of the farms of the poor farmers in the South? Did George know practical ways to make farming more prosperous?

Ans. *Farmers were barely making a living by growing cotton, but this was the only crop they had ever grown. They did not know practical ways to produce more of their own food to eat or to earn enough money from their cotton cash crop. George knew ways to improve the soil and grow healthy plants. He knew ways to prevent insects and other pests. He knew how to pre-pare healthy food that tasted good for families to eat during the winter.*

CHAPTER 9

1. List some of the problems George identified at Tuskegee as he first began to work there.

Ans. *Classrooms, labs, and equipment were not adequate for class-es. There was not enough money to operate the school. Many students could barely read and write. Students lived in extreme poverty.*

2. What were some of the ways in which George was able to make improvements in the school?

Ans. *He was able to get a grant from the Smithsonian and work di-rectly with the federal government. As Tuskegee became more recognized, more funding became available.*

3. In your opinion, why were poor farmers reluctant to stop planting cotton as their only cash crop?

Ans. It was the only cash crop they had ever planted.

4. What was the Farmer's Institute? What were some of the helpful things it accomplished?

Ans. Once a month, farmers and their wives were invited to Tuskegee to learn new farming practices, see demonstrations of practical things they could do to improve their crops, and ask questions about what he was doing. They learned useful practical methods about how to produce better crops. They had an opportunity to bring soil samples for George to analyze and learned how to put nutrients back into the soil. They learned how to get rid of pests invading their crops.

5. What was George's idea to take this information to farmers who did not live near enough to attend the Farmer's Institute?

Ans. George began to take wagons out to farmers who lived too far to take off a day to attend the Farmer's Institute. He demonstrated new farming practices and showed them other things being taught at the Farmer's Institute.

6. How did his idea of going out to farmers in remote areas greatly expand and get funded?

Ans. A banker named Morris Jesup wanted to do something to help poor black farmers in the South. He equipped Jesup Agricultural Wagons with updated information and farming equipment. Thousands of farmers were able to get information that showed them successful farming methods. Later models were carried by pickups and were known as Tuskegee Institute

Movable School and worked with the Alabama Cooperative Extension System.

7. What did Andrew Carnegie do to improve the Institute?

Ans. He donated a library to the school.

8. What other influential people visited Tuskegee to see the work George was doing?

Ans. The president of Harvard University and William Taft, the Secretary of War.

9. Why did a delegate from Africa come to Tuskegee to visit George?

Ans. They wanted to learn how to produce milk from soybeans, since they had difficulty keeping cows alive during the hot drought conditions.

CHAPTER 10

1. How important were the schools that were organized and built for black children? In the early 1900s, were any of the schools in the South integrated where both white and black children attended together?

Ans. They were very important. They taught black children to read and write and do math. These skills helped them prepare for better jobs and opportunities in the future. None of the schools in the early 1900s were integrated. They remained segregated for many years until the Civil Rights Era made changes.

2. What was the purpose of Frances Johnson's trip to south Alabama?

Ans. She wanted to document and photograph improvements in small black schools in the South.

3. Why did some angry white men threaten to kill the young black man who was helping Frances Johnson into the hotel where she had planned to stay?

Ans. They maintained strict segregation between races and thought a black man should never be alone with a white woman.

4. Were George, Frances, and Nelson all able to get back to Tuskegee without any injuries?

Ans. Yes.

5. Several years ago, there was a big news story about an American magazine publisher who had to go into hiding to protect his life. The publisher had printed a political caricature of the leader of an extreme religious sect. Some members of this sect felt it was their duty to kill the publisher whom they believed had insulted their leader. In your opinion, do you think murder or violence is ever justified when a tradition or person is insulted?

Ans. Absolutely not. Add your opinions.

6. What is meant by an idea known as white supremacy? In your opinion, do you think most Americans today have rejected white supremacy?

Ans. A belief that white people belong to a race that is smarter and more superior to black people. It is almost never given approval today among mainstream Americans.

Ans. Most Americans are ashamed that this belief was in their past. Add your opinion.

7. At one time in the past, white supremacist bullies tried to maintain a tight control over what black people could do. What kind of tactics did they use?

Ans. Violence and intimidation and fear.

8. Did the three men who rode up shooting their guns at Nelson cause George to change his mind about how he viewed all white people in general? How did he feel about speaking to white audiences?

Ans. No. He knew there were deep pockets of racism in the South, but he also knew there were also many white people who respected his intelligence and character and sought his advice. It almost never kept him from speaking to white audiences.

9. Did George believe God intended for him to only use his knowledge to help black people?

Ans. He believed God wanted him to use his knowledge and expertise about plants and farming to help poor farmers learn to be prosperous and successful. He did not believe God wanted him to limit his help and knowledge only to black people.

CHAPTER 11

1. Why did George decline several well-paying jobs with good equipment and labs?

Ans. He felt that his God-given purpose was at Tuskegee.

2. Why did George start his day before sunrise in the woods?

Ans. This was an important time for George to hear God's guidance and encouragement.

3. Why did George begin to spend more of his time as a scientific researcher and speaker in addition to his teaching duties at Tuskegee?

Ans. He wanted to discover more healthy, low-cost products that could be made from peanuts, soybeans, and other plants with the hope that businesses would begin to manufacture and sell them. This would give farmers options for cash crops other than cotton and could be a source of food during the winter.

4. What new crop did George correctly predict would become a valuable cash crop in the future?

Ans. Soybeans.

5. How did George convince a Department of Agriculture committee in Washington, D.C. that sweet potato flour was a good, inexpensive source of food?

Ans. When skeptics had doubts that sweet potato flour would be practical, George took the entire committee to a local bakery and made loaves of aromatic, fresh-baked breads from sweet potato flour and served everyone samples.

CHAPTER 12

1. When George received a letter informing him that boll wee-
    vils were headed toward Alabama cotton fields, what crops
    did he recommend that farmers plant and why?

Ans. *He recommended planting legumes, such as peanuts, soy-*
beans, and cow-peas, as well as sweet potatoes. These plants
grew well and replaced many of the nutrients in the soils that
years of planting cotton had depleted. They would also be a
source of food during the winter when the cotton crops didn't
produce another cash crop.

2. For years, many farmers had depended on receiving cash for
    their cotton crops as a way to make a living. In 1904, due to
    the boll weevils, there was no profit from their cotton crops.
    For the farmers who had ventured to add other crops, espe-
    cially peanuts, there was a little food to eat. Did the peanut
    crops provide an adequate source of cash for the farmers that
    year?

Ans. *There was not enough cash to start another crop next spring.*

3. What three things did George do to help the farmers sell their
    new crops for a profit?

Ans. *He first sought God's help in finding answers. He spent several*
days in his lab looking for useful products that could be made
from peanuts and sold for a profit. His search yielded hun-
dreds of useful products that businessmen began to manufac-
ture and sell in markets. He persuaded Congress to put tariffs
on peanuts coming from Asian markets so American farmers
would be able to sell their crops at a fair price.

4. What was George's inspiration for discovering many valuable chemicals and products from peanuts?

Ans. *One morning George sensed God leading him to go to his lab and separate the peanuts into a variety of chemicals; then he recombined these chemicals under different temperatures and pressures to form new products.*

5. As the story about George's inspiration for finding new products from peanuts spread across the campus, what extracurricular study did a group of students ask George to lead? How long did this study continue?

Ans. *Students who were hungry to understand the Bible persuaded George to teach a weekly Sunday night Bible study. The study began in 1907 and lasted for thirty years.*

CHAPTER 13

1. What problem finally convinced many farmers not to rely only on cotton crops to make a living?

Ans. *Boll weevils invaded cotton crops across the South, ruining entire crops. Since cotton was their only crop, there was no money produced the entire year.*

2. What crops other than cotton did farmers begin to grow and rotate from year to year?

Ans. *Peanuts, soybeans, chickpeas, sweet potatoes, and other crops*

3. What did George wish his surrogate father, Moses Carver, could have seen before he died?

Ans. *He wished that Moses Carves could have seen a new generation of farmers who were learning to recycle supplies and equipment in the same way George had learned to recycle from him.*

4. When did America join the Allies in World War I? What were two ways in which George helped contribute to the war effort?

Ans. *America joined the war effort in 1917. George used his research skills to find dyes needed for army uniforms, newspapers, and other products.*

Ans. *He discovered hundreds of plants that could be used to make the needed dyes. He dried fruits and vegetables so that they didn't rot on long trips across the ocean*

5. In the midst of a talk before a Senate committee, one of the senators made an insulting remark to George. How did George respond? In your opinion, what would have been the results if George had responded angrily?

Ans. *Instead of responding to the senator with another insult, his response was a calm humorous remark. Second question is an independent response. Possible outcome is that George could have been dismissed at the end of his allotted ten minutes, and his goal of getting tariffs on Asian peanuts would never have happened.*

6. What did George do to help southern farmers get a fair price for their peanuts in America and not be undercut by cheap peanuts sold by Japanese and Asian farmers? What is a tariff?

Ans. *He discovered many useful products that could be made from peanuts and sold in stores. He persuaded the Department of Agriculture to put tariffs on Asian peanuts, so that American farmers could make enough profit to justify planting peanuts. Tariffs are a form of a federal tax on imports.*

7. What advice did George give to help farmers have more productive crops? Why did some farmers choose not to take his advice?

Ans. *He advised farmers to stop planting cotton on the same soil every year and plant crops that replaced the nutrients in the soil that years of planting cotton had depleted. Many farmers had always planted cotton and understood how to buy seeds, take the picked cotton to the cotton gin to remove the seeds, bail the cotton, and transport the cotton to a market for cash. Some farmers decided to continue doing what they had always done because they were afraid to plant a new crop they were not familiar with.*

8. What was one way George researched peanuts to discover new products?

Ans. *One research method was to separate the peanut into individual components and then re-combine them under different temperatures and pressures to discover new products.*

9. Why did George advise farmers to reduce or abandon growing cotton?

Ans. *He knew that planting cotton year after year on the same soil would deplete the soil of its nutrients and the cotton crops would not grow well.*

CHAPTER 14

1. What were some of the events that increased George's popularity and fame during his first thirty-nine years at Tuskegee?

Ans. *He discovered many useful products that could be made from ordinary plants, and reported on them at conventions, hearings, and other meetings. His discoveries were also reported in national newspapers and magazines, increasing private donations to the school. The library was a donation from Andrew Carnegie who had heard of the work George was doing. The Smithsonian Institution donated funds to George to purchase equipment for the Tuskegee labs. He began the Farmer's Institute where farmers could learn how to improve farming methods, first as monthly meetings at Tuskegee and later from wagons that carried these ideas directly to thousands of farmers throughout Alabama. He worked with businessmen to get the useful products he discovered into stores for profits. His recommendations helped farmers survive the boll weevil invasion and the Great Depression.*

2. Americans learned more about George during a celebration of his fortieth year at Tuskegee. What kind of publications wrote about his life?

Ans. Time *and* Life *magazines and a variety of newspapers.*

3. What was the purpose of the museum that was established a few years later? What did George consider his greatest accomplishment in life?

Ans. It was an encouragement to many people as they saw how a poor, sickly baby born to a slave was able to pursue and accomplish his dream. The opportunity to use his knowledge and expertise to help both black and white farmers in the South become better and more prosperous farmers.

4. In your opinion, do you think George regretted not accepting one of the prestigious job offers that paid a top salary, included an up-to-date laboratory, and held a position of authority? Why or why not?

Ans. He never indicated that he had regrets about his decision to move to Tuskegee or to decline better jobs with higher salaries.

5. Where did George obtain the $32,000 he gave to a George Washington Carver foundation?

Ans. The money was saved over a period of forty years out of his own salary and from frugal living.

6. The year 1904 was a disaster for many southern farmers. What happened in that year and how did that year compare to farms a century later?

Ans. The boll weevil infestation was a disaster in 1904 whereas a century later, these same fields produced thriving healthy plants due to the adoption of improved farming methods such as crop rotation and erosion control measures.

7. What did George do to help farmers make a living with their peanut crops (two ways)?

Ans. He discovered new products made from peanuts that could be sold for profit. And he petitioned congress for tariffs to be added to imported peanuts so that American farmers could realize a fair profit.

8. What happened in 1929 that caused Americans everywhere to worry about having enough food to eat?

Ans. The U.S. had entered into the Great Depression such that many people lost their jobs, banks failed, and farmers had difficulty getting the basic necessities for themselves and crops for others.

9. Name at least four famous people from around the world who were friends with George or who wanted his advice.

Ans. Mahatma Gandhi, Thomas Edison, Henry Ford, Franklin Roosevelt, other presidents, cabinet members, and government leaders.

THE SURROUNDING STORIES

1. Refer to the timeline of events before George W. Carver was born through events that occurred after he died. Copy the timeline on a longer piece of paper to make room for more events. Add other historical events, as well as personal family history if you have access to information about things such as births, deaths, weddings, honors, military service, etc. to get

some perspective on how you and your family members fit into this period of history.

Ans. Individual answers.

2. Find as many things as you can in the story that refers to the year 1859. (Include two years before or after 1859.) Add these events to your original timeline.

Ans. Individual answers.

3. It seemed puzzling to 18th and 19th century Americans why there were great cultural differences between the people of Europe who were the primary settlers of America and the people of Africa, who were brought in as slaves. It was a common belief that the cultures were different because Africans were intellectually inferior. **This was never true!** What were actually the reasons for the **cultural** differences in these groups?

Ans. The isolation of much of Africa essentially separated this population from other nations by geographic barriers, such as the Sahara Desert and the oceans surrounding the continent. This prevented most native Africans from gaining knowledge of the technological advancements of other countries. In addition, there was a prevailing superstitious culture among native Africans.

4. Did slavery originate in America?

Ans. No.

5. What group of Americans owned most of the slaves?

Ans. Plantation owners and owners of big farms used slave labor. Property owners with small farms generally used their own family members to do the work and seldom bought slaves. When Moses and Susan Carver bought George's mother, it was an unusual solution, but one based on kindness and friendship. Everyone understood that Mary would be free to leave as soon as the Civil War was over.

6.   How did the invention of the cotton gin affect slavery?

Ans.   *The cotton gin enabled seeds to be removed from the cotton much more quickly than could be done by hand. The gin resulted in plantation owners greatly expanding their cotton fields and thus the need for more slavery to work the fields.*

7.   How many slaves were living in America by the end of the Civil War?

Ans.   *After the Revolutionary War, there were 700,000 slaves living in the U.S. After the Civil War, this number had increased to almost 4,000,000.*

8.   By the year 1904, many southern farmers were facing a disaster with their cotton crops. What happened in that year and how did that year compare to farms 100 years later?

Ans.   *An invasion of boll weevils had ruined most of the cotton crops. Farmers had no crops to sell for cash and they had not planted many food crops.*

Many people struggled to survive. 100 years later, farmers were practicing good farming techniques. Most modern farms are not limited to only one cash crop.

9. Planting nothing but cotton year after year caused many problems for poor southern farmers when George came to Tuskegee. What problems did this practice cause?

Ans. Years of only planting cotton depleted nitrogen and other nutrients from the soil, resulting in unhealthy plants with small yields of crops.

10. What kinds of changes did George recommend for farmers to remedy the problems?

Ans. Plant peanuts, chickpeas, sweet potatoes, and other plants that produce nitrogen in their roots. Then plow the roots under the next year to put nitrogen back into the soil. Practice crop rotation and erosion control to restore the soil to a healthy state.

11. What did George do to help farmers make a living with their peanut crops? (Two ways.)

Ans. He researched peanuts and other plants to find ways they could be used to produce useful products. He met with Congressional committees to ask them to put tariffs on foreign competitors that were selling peanuts at prices too cheap to compete and make a profit.

12. George was born during the Civil War. What were the conditions in the South after the war ended? Why were the northern states not affected in the same ways?

Ans. Conditions were widespread infrastructure damage to communities. Livestock and crops were often stolen or destroyed by the fighting.

Battles were mostly fought in the South.

13. George also lived during World War I. What were some of the ways in which he supported the war for the U.S.?

Ans. *After the war began, aniline dyes that were only available from Germany were quarantined. George researched many plants and discovered new substitutes for the dyes needed for making printer's ink used to publish newspapers. The dyes were also necessary for making dyes for army uniforms and for making paints for ships and other equipment. He also found ways to dry fruits and vegetables, preventing their decay during long trips across the ocean. And he invented a method for making flour for bread-making from sweet potatoes that could be used on long trips.*

14. George died in 1943 as World War II was underway, and America was recovering from the Great Depression. What was the Great Depression? What were some of the ways George had been able to help farmers be prepared for this time?

Ans. *The Great Depression began in 1929. Millions of people lost their jobs and did not have enough money to buy the basic things they needed. Banks failed and could not loan money to people. George introduced cash crops other than cotton and researched ways the new crops could be sold for profit.*

15. In your opinion, why were actions against slavery not immediately proposed by the First U.S. Congress, even when most of the writers of the Constitution did not approve of slavery?

Ans. *Most of the writers of the Constitution agreed with President Washington and looked for ways to end slavery gradually. They were afraid the former slaves would not be able to find adequate food, homes, and incomes. They were afraid several states would withdraw from the United States.*

16. Was the idea that some races were more superior than other races a widespread belief during the time of slavery?

Ans. *Yes.*

17. Compare the speech by William Lloyd Garrison with the one given by Alexander Stephens. Which man believed slaves were fully human, made in the image of God? Which man believed slaves were from an inferior race? Which man was an abolitionist? Which man believed in "white supremacy"?

In a fiery speech in 1854, William Lloyd Garrison said, "But, if they [the slaves] are men, if they are to run the same career of immortality with ourselves … then when I claim for them all that we claim for ourselves, because we are created in the image of God, I am guilty of no extravagance, but am bound, by every principle of honor, by all the claims of human nature, by obedience to Almighty God, to 'remember them that are in bonds as bound with them,' and to demand their immediate and unconditional emancipation."
In a speech given in 1861 by the vice president of the new Confederacy, Alexander Stevens said, "Our new government is founded upon … the great truth that the Negro is not equal to the white man; that slavery, subordination to the superior race, is his natural and normal condition."

Ans. *William Lloyd Garrison was an abolitionist who tried to help free all the slaves at once. He believed all humans were created in the image of God. Alexander Stevens was a white supremacist who wanted the slaves to remain as slaves. Stevens believed Negroes were inferior to the white man and that slavery was the natural and normal condition of the slaves.*

18. At the same time black citizens in America were victims of racists beliefs, Jewish citizens in Germany were victims of anti-Semitic racist beliefs. Give an example of how the civil rights of both black Africans in America and Jews in Germany were victims of racist beliefs. What was the lie behind these actions?

Ans. Both Jews and African Americans experienced racial discrimination. They were restricted in where they could live and work and could not legally marry someone from a "superior race."

19. When George transferred to Iowa State University, he was the only black student at the college. He was treated differently than the other students by fellow classmates and by teachers. Look up the word "stereotype" and tell if you think George was being judged by stereotypic beliefs because of his dark skin instead of by his individual character and abilities.

Ans. Your opinion.

20. Who was Martin Luther King Jr.? Did he promote violent protests or non-violent protests?

Ans. Martin Luther King Jr. was a southern pastor who led thousands of African Americans to protest discriminatory practices regarding employment and other rights. He was responsible for many changes in the South regarding racial discrimination through his non-violent protests.

21. What did the Civil Rights Law of 1964 accomplish?

Ans. *It outlawed discrimination based on race, color, religion, sex, and national origin.*

22. Some people are promoting racial curricula in schools, such as Critical Race Theory and the 1619 Project to show that racism and white supremacy are still major evils of American culture. In your opinion, is the following statement accurate? Explain your answer. "Although slavery and racism were a part of American history, they were not born in America. However, they did die here." (Quote by unknown author.)

Ans. *Your opinion.*

23. What does Dr. Alveda King mean by the phrase "one blood" as used in Acts 17:26?

Ans. *She believed that all humans were descendants of the original first parents, Adam and Eve, who were designed and created by God, and did not gradually evolve from ape-like animals. Every human on earth could trace their ancestors back to Adam and Eve.*

24. What did Hitler believe about different races? Where do you think he got his extreme ideas that led him to order the murder of millions of Jewish people he considered "inferior"?

Ans. *He believed that all humans evolved from animals and that during the evolutionary process, some humans evolved into inferior races. He believed some humans evolved into superior races. He did not believe humans descended from Adam and Eve, the original human set of parents, whom God designed and made in His image.*

25. Choose one of the following: Write a possible radio interview with a character in the story. Use one person to be the interviewer and one person to be the character being interviewed.

Ans. Individual answers.

PARTIAL LIST OF REFERENCES

Information was obtained for this book from the helpful staff at Tuskegee University and the George Washington Carver Museum, as well as from the George Washington Carver National Memorial, which was dedicated to him at his birth place in Diamond, Missouri.

Benge, Janet & Geoff. *George Washington Carver: From Slave to Scientist*. (Lynnwood, WA: Emerald Books, 2001.)

Federer, William, J. *George Washington Carver: His Life and Faith in His Own Words*. (Amerisearch, Inc., 2002.)

McKay, John P., Hill, Bennett D., Buckler, John. *A History of Western Society, Volume I: From Antiquity to the Enlightenment*. (Boston: Houghton Mifflin Company. 1995.)

Weikart, Richard. *The Death of Humanity and the Case for Life*. (Washington, DC: Regency Faith Publishing. 2016.)

ACKNOWLEDGMENTS

As with any book, many people were part of this project. I would like to name a few:

- Charli Kendricks and Rose Carman for their talented drawings
- Tracy Crump for advising and editing various sections of this story
- Faithe Thomas for steering me through numerous requirements, recommendations, and optional decisions
- Marni Kendricks for patiently helping me when I accidentally lost sections of the manuscript

www.ingramcontent.com/pod-product-compliance
Lightning Source LLC
Chambersburg PA
CBHW071425210326
41597CB00020B/3661